Monographs in

Volume 5

Hecke's Theory of Modular Forms and Dirichlet Series

2nd Printing with Revisions

Monographs in

Volume 5

Hecke's Theory of Modular Forms and Dirichlet Series

2nd Printing with Revisions

Bruce C Berndt
University of Illinois at Urbana-Champaign, USA

Marvin I Knopp
Temple University, USA

NEW JERSEY • LONDON • SINGAPORE • BEIJING • SHANGHAI • HONG KONG • TAIPEI • CHENNAI

Published by

World Scientific Publishing Co. Pte. Ltd.
5 Toh Tuck Link, Singapore 596224
USA office: 27 Warren Street, Suite 401-402, Hackensack, NJ 07601
UK office: 57 Shelton Street, Covent Garden, London WC2H 9HE

British Library Cataloguing-in-Publication Data
A catalogue record for this book is available from the British Library.

Photo of Erich Hecke (page vi) courtesy of Vandenhoeck & Ruprecht.

HECKE'S THEORY OF MODULAR FORMS AND DIRICHLET SERIES

ISBN-13 978-981-270-635-5
ISBN-10 981-270-635-6

Printed in Singapore.

In Memory of Hans Rademacher, the Father of our Mathematical Family

Erich Hecke

Preface in Two Acts with a Prelude, Interlude, and Postlude

Prelude

Thirty-seven years have elapsed between the first version and the present version of this monograph. We begin with the first author's slightly edited preface from his first version. We then provide a lengthier second preface composed by the second author.

The Original Preface

These notes are part of a course on modular forms and applications to analytic number theory given by the first author at the University of Illinois at Urbana-Champaign in the spring of 1970. The existing accounts [47], [48], [87] of Hecke's theory of modular forms and Dirichlet series are somewhat concise. Therefore, it has been our intention to present a more detailed account of a major portion of this material for those who are unfamiliar with this beautiful theory. Readers already familiar with Hecke's theory will find little that is new here.

The first author is especially grateful to Ronald J. Evans for providing a new proof of a fundamental region for Hecke's modular groups, which we present here. We express our thanks also to Elmer Hayashi for a detailed reading of the manuscript and to Harold Diamond for several suggestions.

Bruce Berndt, May, 1970 & May, 2007

Interlude

The first author mailed a copy of his notes on Hecke's theory of modular forms and Dirichlet series to Dr. Jürgen Elstrodt, who at that time was at Universität München. He responded with about a dozen pages of detailed comments, which, after an undeservedly quick reading, were deposited in the first author's file cabinet for approximately thirty-five years, until they were dusted off and sent to the second author for incorporation in the new version. We hope that it is not too late to thank Elstrodt for his kind suggestions and patience.

The Second Preface

In the spring of 1971, I received the following letter, dated June 17. Since it is brief, I quote it in full.

> Under separate cover, I am sending you a copy of some lecture notes, "Hecke's theory of modular forms and Dirichlet series." I would appreciate any comments, corrections, criticisms, or suggestions that you may have. Thank you very much.
>
> Most sincerely, (signed) Bruce

To establish the context of this letter, I recall that in the spring of 1938 Erich Hecke gave an important series of lectures at the Institute for Advanced Study, Princeton, on his correspondence theory published in 1936. The notes from these lectures, taken by Hyman Serbin and produced in planographed form by Edwards Brothers of Ann Arbor, received only limited circulation. To my knowledge there are only a few copies extant in mathematics libraries (for example, the University of Illinois at Urbana-Champaign) and private collections of professional mathematicians.

In 1970 Berndt produced a set of lecture notes based upon Hecke's notes, but with the addition of many details omitted from Hecke's original notes. The more extensive notes, too, had only limited circulation.

For the past thirty-five years I have employed both sets of notes to introduce graduate students to the Hecke theory and the broader theory of modular/automorphic forms. During this time my Ph.D. students and others frequently asked why Berndt's notes had never been published. Because we are convinced that the reactions of these students reflect a genuine usefulness of these notes to the mathematical community, we have undertaken the task of publishing this book based upon them, corrected and modified where necessary, and expanded to include some of the many new developments in the theory during the past decades, as well as relevant earlier work not previously included. We stress that the Hecke correspondence theory has remained an active feature of research in number theory since the 1930s and, in fact, its importance is perhaps better understood today than it was in 1936.

The first six chapters of this book follow the organization of Berndt's original notes, hence that of the first part of Hecke's notes as well. Beyond this, we have added two completely new chapters based upon work done since 1970 and upon earlier work not originally understood to lie within the circle of ideas surrounding Hecke's correspondence theorem.

Chapter 7 features Bochner's important generalization of Hecke's correspondence theorem and some closely related results. Chapter 8 is devoted to the great variety of identities related to the Hecke correspondence theory (but not explicitly present in that theory) that have been developed over the years. Among others, these identities are due to S. Ramanujan, N. S. Koshliakov, G. N. Watson, A. P. Guinand, K. Chandrasekharan, R. Narasimhan, and Berndt. Some antedate Hecke's work, while others are more recent.

Marvin Knopp, April, 2007

Postlude

We are grateful for the comments made by our students over the past several decades. More recently, Shigeru Kanemitsu and Yoshio Tanigawa offered several additional remarks and references. We thank Hilda Britt for expertly typing most of our manuscript and Tim Huber for his graphical expertise.

Contents

Chapter 1

Introduction

The classical theta function, defined for $\operatorname{Im}\tau > 0$ by

$$\theta(\tau) = \sum_{n=-\infty}^{\infty} e^{\pi i n^2 \tau},$$

satisfies the modular transformation law

$$\theta(-1/\tau) = (\tau/i)^{\frac{1}{2}}\theta(\tau). \tag{1.1}$$

Perhaps the best-known way to derive the functional equation of the Riemann zeta function $\zeta(s)$,

$$\pi^{-s/2}\Gamma(s/2)\zeta(s) = \pi^{(s-1)/2}\Gamma(\{1-s\}/2)\zeta(1-s), \tag{1.2}$$

is by way of (1.1) [107, p. 22]. Conversely, it is not difficult to show that (1.2) implies (1.1), but this derivation requires the use of the Phragmén-Lindelöf Theorem [105, §5.65]. In 1921 H. Hamburger [38] showed that, under certain auxiliary analytic conditions, $\zeta(s)$ is essentially the only solution to the functional equation (1.2). For a more transparent proof, see C. L. Siegel's paper [99], [100, pp. 154–156]. More specifically, they proved that if $f(s)$ is a Dirichlet series satisfying the aforementioned auxiliary restrictions, and if

$$R(s) = \pi^{-s}\Gamma(s)f(2s), \qquad R(s) = R\left(\frac{1}{2} - s\right), \tag{1.3}$$

then $f(s)$ is a constant multiple of $\zeta(s)$. See also [107, pp. 31–32]. That $f(2s)$, as opposed to $f(s)$, appears in (1.3) guarantees *a priori* that the inverse Mellin transform of $R(s)$, an exponential series, has its coefficient sequence supported on integral squares, and thus it has the general shape of

$\theta(\tau)-1$. The proof of Hamburger's Theorem is then completed by showing that this inverse Mellin transform is in fact a constant multiple of $\theta(\tau)-1$.

Of even greater interest within the context of the present work is a second, distinct version, due to Hecke, of the Hamburger theorem. This version is, in fact, a direct consequence of a general correspondence theorem proved by Hecke in 1936 [47] (the "main correspondence theorem" of Chapter 2, below) and the fact that, under certain conditions of regularity, $\theta(\tau)$ is the only solution to (1.1) that is periodic (with period 2). For further details concerning the two formulations of Hamburger's theorem, see the introduction to Hecke's final published paper [49], [62, esp. pp. 201–207], and the **Application** following Remark 7.4.

Throughout the sequel we let $\tau = x + iy$ and $s = \sigma + it$ with x, y, σ, and t real. We denote the upper half-plane, $\{\tau : y > 0\}$, by $\mathcal{H}$. The set of all complex numbers will be denoted by $\mathbb{C}$, the set of all real numbers by $\mathbb{R}$, the set of all rational numbers by $\mathbb{Q}$, and the set of all rational integers by $\mathbb{Z}$. We adopt the following *argument convention*: for $w \in \mathbb{C}$, $w \neq 0$, and $k \in \mathbb{R}$, w^k is defined by

$$w^k = |w|^k e^{ik \arg w}, \qquad -\pi \leq \arg w < \pi. \tag{1.4}$$

The summation sign $\sum$ with no indices always means $\sum\limits_{n=1}^{\infty}$. We write $\int_{(c)}$ for $\int_{c-i\infty}^{c+i\infty}$, where c is real and the path of integration is the straight line from $c-i\infty$ to $c+i\infty$. We often use the symbol A to denote a positive constant, not necessarily the same with each occurrence.

Chapter 2

The main correspondence theorem

Before proving the main theorem we first establish a couple of lemmas.

Lemma 2.1. *Let $\varphi(s) = \sum a_n n^{-s}$. Then, $\varphi(s)$ converges in some half-plane if and only if $a_n = O(n^c)$, for some constant c, as n tends to ∞.*

Proof. First, assume that $a_n = O(n^c)$ as n tends to ∞. Let $\sigma \geq c+1+\epsilon$ for some constant $\epsilon > 0$. Then,

$$\left|\sum a_n n^{-s}\right| \leq A \sum n^{c-\sigma} \leq A \sum n^{-1-\epsilon} < \infty.$$

Therefore, $\varphi(s)$ converges for $\sigma \geq c+1+\epsilon$.

Conversely, if $\varphi(s)$ converges for $s = s_0 = \sigma_0 + it_0$, then $a_n n^{-s_0}$ tends to 0 as n tends to ∞. In particular, $a_n = O(n^{\sigma_0})$. □

Lemma 2.2. *Let $\lambda > 0$ and $c \geq 0$. Suppose that*

$$f(\tau) = \sum_{n=0}^{\infty} a_n e^{2\pi i n \tau/\lambda}$$

is analytic on $\mathcal{H}$.

(i) *If $a_n = O(n^c)$, then $f(\tau) = O(y^{-c-1})$, uniformly for all x, $-\infty < x < \infty$, as $y \to 0+$.*

(ii) *If $f(\tau) = O(y^{-c})$ as $y > 0$ tends to 0, uniformly for all x, then $a_n = O(n^c)$ as n tends to ∞.*

Proof of (i). For $u \geq 0$, $u^c \exp(-2\pi uy/\lambda)$ achieves its maximum at $u = c\lambda/2\pi y$. Thus, as $y > 0$ tends to 0,

$$\begin{aligned}
|f(\tau)| &\leq \sum_{n=0}^{\infty} |a_n| e^{-2\pi ny/\lambda} \\
&\leq A \sum n^c e^{-2\pi ny/\lambda} \\
&\leq A \left(\int_0^{c\lambda/2\pi y} u^c du + \int_{c\lambda/2\pi y}^{\infty} u^c e^{-2\pi uy/\lambda} du \right) \\
&\quad + O(\{c\lambda/2\pi y\}^c e^{-2\pi(c\lambda/2\pi y)y/\lambda}) \\
&= O(y^{-c-1}) + O\left(y^{-c-1} \int_0^{\infty} u^c e^{-u} du \right) + O(y^{-c}) \\
&= O(y^{-c-1}).
\end{aligned}$$

□

Proof of (ii). Let $\tau_0 \in \mathcal{H}$. Then by Fourier's formula for the coefficients of a Fourier series,

$$\begin{aligned}
a_n &= \frac{1}{\lambda} \int_{\tau_0}^{\tau_0+\lambda} f(\tau) e^{-2\pi i n\tau/\lambda} d\tau \\
&= O\left(\int_{\tau_0}^{\tau_0+\lambda} y^{-c} e^{2\pi ny/\lambda} |d\tau| \right),
\end{aligned}$$

as $y > 0$ tends to 0. If we set $y = 1/n$, we find that $a_n = O(n^c)$ as n tends to ∞. □

Theorem 2.1. *Let $\{a_n\}$ and $\{b_n\}$, $0 \leq n < \infty$, be sequences of complex numbers such that a_n, $b_n = O(n^c)$, as n tends to ∞, for some $c \geq 0$. Let $\lambda > 0$, $k \in \mathbb{R}$, and $\gamma \in \mathbb{C}$. For $\sigma > c+1$, put*

$$\varphi(s) = \sum a_n n^{-s} \quad \text{and} \quad \psi(s) = \sum b_n n^{-s}.$$

Define, for $\sigma > c+1$,

$$\Phi(s) = (2\pi/\lambda)^{-s}\Gamma(s)\varphi(s) \quad \text{and} \quad \Psi(s) = (2\pi/\lambda)^{-s}\Gamma(s)\psi(s).$$

For $\tau \in \mathcal{H}$, let

$$f(\tau) = \sum_{n=0}^{\infty} a_n e^{2\pi i n\tau/\lambda} \quad \text{and} \quad g(\tau) = \sum_{n=0}^{\infty} b_n e^{2\pi i n\tau/\lambda}.$$

Then the following two assertions are equivalent.

(i) $f(\tau) = \gamma(\tau/i)^{-k} g(-1/\tau)$.

(ii) $\Phi(s) + a_0/s + \gamma b_0/(k-s)$ *has an analytic continuation to the entire complex plane that is entire and bounded in every vertical strip. Moreover,*

$$\Phi(s) = \gamma \Psi(k-s). \tag{2.1}$$

Remark 2.1. Our formulation of Theorem 2.1 deviates from Hecke's original statement [47], [48] of his correspondence theorem in two ways. In Hecke's work there are not two, but only a single Dirichlet series; that is, $\psi(s) = \varphi(s)$. Also, our boundedness condition in (ii) replaces a corresponding hypothesis of Hecke, who assumes that $(s-k)\varphi(s)$ is an entire function of finite genus, that is to say, there exists an $M > 0$ such that $|(s-k)\varphi(s)| \le \exp\{|s|^M\}$, for all s in $\mathbb{C}$. In Chapter 7 we discuss the extent to which the Dirichlet series $\varphi(s)$ and $\psi(s)$ can differ, and how this difference affects the theory developed in our Chapters 4–6.

We turn to the matter of the differing conditions on boundedness. Clearly, they are equivalent within the framework of the Hecke correspondence theorem. For, condition (ii) is equivalent to (i) in our Theorem 2.1, while the original Hecke version of (ii) (assuming $(s-k)\varphi(s)$ is entire of finite genus) is likewise equivalent to (i) [47], [48]. On the other hand, these conditions on boundedness are not equivalent outside of the context of the correspondence theorem. To see this, recall that

$$\Phi(s) = (2\pi/\lambda)^{-s}\Gamma(s)\varphi(s)$$

and that $\Gamma(s)$ has the following growth properties:

(a) $1/\Gamma(s)$ is entire of finite genus;

(b) $\Gamma(s)$ is bounded in vertical strips (truncated when necessary to avoid the poles of $\Gamma(s)$).

Using these facts, we can reduce the equivalence of the two boundedness assumptions to: $h(s)$ is bounded in vertical strips if and only if $h(s)$ is of finite genus. But this equivalence fails in both directions, since

(c) $h_1(s) = \exp(e^s)$ is bounded in vertical strips, but not of finite genus;

(d) $h_2(s) = \exp(-s^2)$ is of finite genus, but not bounded in vertical strips.

Proof of Theorem 2.1. First assume that (i) is valid. From Euler's integral representation of the Γ-function, for $\sigma > c+1$,

$$\begin{aligned}\Phi(s) &= \sum a_n \int_0^\infty (2\pi n/\lambda)^{-s} u^{s-1} e^{-u} du \\ &= \sum a_n \int_0^\infty u^{s-1} e^{-2\pi n u/\lambda} du.\end{aligned}$$

Since $\sigma > c+1$, we may invert the order of summation and integration by absolute convergence to obtain

$$\begin{aligned}\Phi(s) &= \int_0^\infty u^{s-1} \sum a_n e^{-2\pi n u/\lambda} du \\ &= \int_0^\infty u^{s-1}(f(iu) - a_0) du \\ &= \left\{\int_0^1 + \int_1^\infty\right\} u^{s-1}(f(iu) - a_0) du \\ &= \int_1^\infty u^{-s-1} f(i/u) du - a_0/s + \int_1^\infty u^{s-1}(f(iu) - a_0) du.\end{aligned}$$

Using (i), we find that, for $\sigma > \sup(c+1, k)$,

$$\begin{aligned}\Phi(s) = \gamma \int_1^\infty u^{-s-1+k}(g(iu) - b_0) du + \int_1^\infty u^{s-1}(f(iu) - a_0) du \\ - a_0/s - \gamma b_0/(k-s).\end{aligned} \tag{2.2}$$

Since $f(iu) - a_0$, $g(iu) - b_0 = O(\exp\{-2\pi u/\lambda\})$ as u tends to ∞, it follows by analytic continuation that $\Phi(s) + a_0/s + \gamma b_0/(k-s)$ is an entire function. It is easily seen that $\Phi(s) + a_0/s + \gamma b_0/(k-s)$ is bounded in every vertical strip by taking absolute values in (2.2). If we replace s by $k-s$ in (2.2) and use a formula analogous to (2.2) for $\Psi(s)$, we immediately find that $\Phi(k-s) = \gamma\Psi(s)$.

Conversely, we now assume (ii). By the Cahen-Mellin formula [75, pp. 97–98], for x, $d > 0$,

$$e^{-x} = \frac{1}{2\pi i}\int_{(d)} \Gamma(s) x^{-s} ds. \tag{2.3}$$

Upon letting $x = 2\pi n y/\lambda$ with n, $y > 0$, we find that

$$e^{-2\pi n y/\lambda} = \frac{1}{2\pi i}\int_{(d)} \Gamma(s)(2\pi n y/\lambda)^{-s} ds.$$

Multiplying both sides by a_n and summing on n, we deduce that, for $d > c+1$,

$$f(iy) - a_0 = \sum a_n \frac{1}{2\pi i} \int_{(d)} \Gamma(s)(2\pi ny/\lambda)^{-s} ds \tag{2.4}$$
$$= \frac{1}{2\pi i} \int_{(d)} \Phi(s) y^{-s} ds,$$

where the inversion in order of summation and integration is justified by the absolute and uniform convergence of $\varphi(s)$ on the line $\sigma = d$.

We next move the path of integration to the line $\sigma = -d$. We shall do this by integrating around a rectangle with vertices $\pm d \pm iT$, $T > 0$, applying the residue theorem, and then showing that the integrals along the horizontal sides tend to 0 as T tends to ∞. Now, by Stirling's formula [26, p. 224],

$$|\Gamma(\sigma + it)| \sim (2\pi)^{\frac{1}{2}} |t|^{\sigma - \frac{1}{2}} e^{-\pi |t|/2}, \tag{2.5}$$

as $|t|$ tends to ∞, uniformly on any fixed interval, $\sigma_1 \le \sigma \le \sigma_2$. By hypothesis, $\Phi(s)$ is bounded in every vertical strip. It follows that

$$\varphi(s) = O(|t|^{\frac{1}{2} - \sigma} e^{\pi |t|/2}), \tag{2.6}$$

as $|t|$ tends to ∞, uniformly for $-d \le \sigma \le d$. On the line $\sigma = d$, clearly,

$$\varphi(s) = O(1), \tag{2.7}$$

as $|t|$ tends to ∞. From (2.1) and (2.5), we find that, on the line $\sigma = -d$,

$$\varphi(s) = O(\Gamma(k-s)\psi(k-s)/\Gamma(s)) = O(|t|^{k+2d}), \tag{2.8}$$

as $|t|$ tends to ∞, since $\psi(k-s) = O(1)$. Thus, from (2.6)–(2.8), we see that the hypotheses of the Phragmén-Lindelöf Theorem for a vertical strip [105, p. 180] are satisfied. Thus,

$$\varphi(s) = O(|t|^A), \tag{2.9}$$

uniformly on $-d \le \sigma \le d$. Hence, from (2.5) and (2.9) it is easily seen that the integrals on the horizontal sides approach 0 as T tends to ∞. Therefore, we have

$$f(iy) - a_0 = \frac{1}{2\pi i} \int_{(-d)} \Phi(s) y^{-s} ds - a_0 + \gamma b_0 y^{-k}.$$

Using (2.1), replacing s by $k-s$, and employing a representation for $g(iy)-b_0$ analogous to that of (2.4), we find that

$$\begin{aligned} f(iy) &= \frac{\gamma}{2\pi i}\int_{(-d)} \Psi(k-s)y^{-s}ds + \gamma b_0 y^{-k} \\ &= \frac{\gamma}{2\pi i}\int_{(k+d)} \Psi(s)y^{s-k}ds + \gamma b_0 y^{-k} \\ &= \gamma y^{-k}(g(i/y) - b_0) + \gamma b_0 y^{-k} \\ &= \gamma y^{-k} g(i/y). \end{aligned}$$

Part (i) now easily follows by analytic continuation. This completes the proof. □

Note that $\varphi(s)$ is entire if and only if $b_0 = 0$. If $b_0 \neq 0$, $\varphi(s)$ has a simple pole at $s = k$. Alternatively, one can say that $\varphi(s)$ is entire if and only if $f(\tau)$ vanishes at $i\infty$.

Theorem 2.1 has been generalized by several authors [7], [17], [20, pp. 665–696], [23], [77]. See Chapter 7 for more on S. Bochner's generalization in [17].

In [23], K. Chandrasekharan and R. Narasimhan extended Theorem 2.1 to the broader context of generalized Dirichlet series (as did Bochner in [17], [20, pp. 665–696]), that is, series of the form $\sum a_n \lambda_n^{-s}$, where $\{\lambda_n\}$ is a monotone sequence of positive real numbers such that $\lambda_n \to +\infty$ as $n \to +\infty$. Their version of the fundamental equation (2.1) retains the single factor $\Gamma(s)$, but weakens the restriction on the poles of the (generalized) Dirichlet series. Furthermore, the authors obtain a third relation equivalent to the two assertions concerning $\{a_n\}$ in their generalization of Theorem 2.1. This third identity expresses a sum of the form $\sum\limits_{n\leq x} a_n(x-\lambda_n)^q$ in terms of Bessel functions.

In [7] Berndt extended the results of [23] to the case in which the functional equation analogous to (2.1) contains $\Gamma^m(s)$, where m is any positive integer. As in [23], the author [7] proved the equivalence of three identities. Chapter 8 is devoted to further identities equivalent to the identities of Theorem 2.1.

H. Maass [77, Theorem 35, p. 228] reformulated Theorem 2.1 to obtain a theorem appropriate to the context of nonholomorphic (real analytic) exponential series. The result here is, on its face, far more complex than Theorem 2.1, but in the end the technique of proof is the same, once Maass derives suitable necessary conditions for the Fourier coefficients of his exponential series.

The best-known, most widely applied generalization of Theorem 2.1 is that of A. Weil [112], [114, pp. 165–172], who introduced the notion of "twists" by Dirichlet characters, of Dirichlet series, to carry over the Hecke correspondence theorem to the context of the Hecke congruence subgroups $\Gamma_0(N)$ of the full modular group $SL(2,\mathbb{Z})$. (For $N \in \mathbb{Z}$, $N > 0$, $\Gamma_0(N)$ is the collection of matrices $\left(\begin{smallmatrix} a & b \\ c & d \end{smallmatrix}\right)$ in $SL(2,\mathbb{Z})$ such that $c \equiv 0 \pmod{N}$.) Note that $SL(2,\mathbb{Z})$ is denoted by $G(1)$ in Chapter 5; of course, in the present notation, $SL(2,\mathbb{Z}) = \Gamma_0(1)$ as well. Readers are encouraged to read a second paper by Weil [113], [114, pp. 405–412] in which he applies the Hecke correspondence to problems raised by the evaluation of periods of certain Abelian integrals.

We have already observed that the Riemann zeta-function satisfies the functional equation (1.2), which is of the form (2.1). There are several other well-known Dirichlet series satisfying a functional equation of the form (2.1).

Example 2.1.

1. Let $Q(n_1, \ldots, n_m)$ denote a positive definite quadratic form in m variables with real coefficients and discriminant $d > 0$. The Epstein zeta-function $Z(s,Q)$ is defined for $\sigma > m/2$ by

$$Z(s,Q) = \sideset{}{'}\sum_{n_1,\ldots,n_m=-\infty}^{\infty} \{Q(n_1,\ldots,n_m)\}^{-s},$$

where the prime $'$ indicates the term with $n_1 = \cdots = n_m = 0$ is omitted from the summation. The function $Z(s,Q)$ has an analytic continuation to the entire complex plane and satisfies the functional equation [32]

$$\pi^{-s}\Gamma(s)Z(s,Q) = d^{-\frac{1}{2}}\pi^{s-\frac{1}{2}m}\Gamma(\tfrac{1}{2}m - s)Z(\tfrac{1}{2}m - s, Q^{-1}),$$

where Q^{-1} denotes the inverse of Q.

2. Let K denote an algebraic number field of degree $n = r_1 + 2r_2$, where r_1 denotes the number of real conjugates of K and $2r_2$ the number of complex conjugates of K. Then the Dedekind zeta-function of K is defined for $\sigma > 1$ by

$$\zeta_K(s) = \sum F(m)m^{-s},$$

where $F(m)$ denotes the number of nonzero integral ideals of norm m in K. Let

$$B = 2^{-r_2}\pi^{-\frac{1}{2}n}|d|^{\frac{1}{2}},$$

where d is the discriminant of K, and put

$$\Phi(s) = B^s\{\Gamma(\tfrac{1}{2}s)\}^{r_1}\{\Gamma(s)\}^{r_2}\zeta_K(s).$$

Then, $\zeta_K(s)$ has an analytic continuation to the entire complex plane and satisfies the functional equation [70, p. 67]

$$\Phi(s) = \Phi(1-s).$$

Note that $\Phi(s)$ has the same form as the analogous function in Theorem 2.1 if and only if K is an imaginary quadratic field, that is, $r_1 = 0$ and $r_2 = 1$.

3. Next, let χ be a nonprincipal, primitive character $\pmod{k}$. For $\sigma > 0$, the Dirichlet L-function is defined by

$$L(s,\chi) = \sum \chi(n)n^{-s}.$$

Let

$$\Phi(s,\chi) = (\pi/k)^{-\frac{1}{2}(s+a)}\Gamma\left(\tfrac{1}{2}\{s+a\}\right)L(s,\chi),$$

where $a = 0$ if $\chi(-1) = 1$ and $a = 1$ if $\chi(-1) = -1$. Then $L(s,\chi)$ can be continued to an entire function satisfying the functional equation

$$\Phi(s,\chi) = \epsilon(\chi)\Phi(1-s,\bar{\chi}),$$

where $|\epsilon(\chi)| = 1$ [2, p. 371], [27, pp. 71–72].

4. Ramanujan's tau-function $\tau(n)$ is defined by

$$q\prod_{n=1}^{\infty}(1-q^n)^{24} = \sum \tau(n)q^n, \quad |q| < 1.$$

It can be shown that

$$f(s) = \sum \tau(n)n^{-s}$$

converges absolutely for $\sigma > 13/2$, can be analytically continued to an entire function, and satisfies the functional equation [43, Chapter 10]

$$(2\pi)^{-s}\Gamma(s)f(s) = (2\pi)^{s-12}\Gamma(12-s)f(12-s).$$

Definition 2.1. Let $\varphi = \psi$ and suppose that φ satisfies condition (ii) of Theorem 2.1. Then, we say that φ *has signature* (λ, k, γ).

Note that if $\varphi \not\equiv 0$ has signature (λ, k, γ), then $\gamma = \pm 1$. For, combining the relations

$$\Phi(s) = \gamma\Phi(k - s)$$

and

$$\Phi(k - s) = \gamma\Phi(s),$$

we deduce that

$$\Phi(s) = \gamma^2\Phi(s).$$

Some of the series discussed above yield examples of Dirichlet series of signature (λ, k, γ). Thus, $\zeta(2s)$ has signature $(2, \frac{1}{2}, 1)$. If $Q(n_1, \ldots, n_m) = n_1^2 + \cdots + n_m^2$, then

$$\zeta(s, Q) = \zeta_m(s) = \sum r_m(n)n^{-s}$$

has signature $(2, \frac{1}{2}m, 1)$. Here, $r_m(n)$ is the number of integral vectors $\{n_1, \ldots, n_m\}$ such that $n = n_1^2 + \cdots + n_m^2$. Alternatively, $r_m(n)$ denotes the number of representations of the positive integer n as the sum of m squares. If K is an imaginary quadratic field, then $\zeta_K(s)$ has signature $(|d|^{\frac{1}{2}}, 1, 1)$. Ramanujan's Dirichlet series $f(s) = \sum \tau(n)n^{-s}$ has signature $(1, 12, 1)$.

Definition 2.2. We say that f belongs to the space $M(\lambda, k, \gamma)$ if

(i) $f(\tau) = \sum_{n=0}^{\infty} a_n e^{2\pi i n\tau/\lambda}$,

where $\lambda > 0$ and $\tau \in \mathcal{H}$, and

(ii) $f(-1/\tau) = \gamma(\tau/i)^k f(\tau)$,

where $k > 0$ and $\gamma = \pm 1$.

We say that f belongs to the space $M_0(\lambda, k, \gamma)$ if f satisfies conditions (i) and (ii), and if

(iii) $a_n = O(n^c)$,

for some real number c, as n tends to ∞.

It is a direct consequence of (ii) that $\gamma = 1$ if $f(i) \neq 0$. If $f(\tau) \not\equiv 0$, the convention (1.4) implies that $\gamma^2 = 1$, since $\tau \in \mathcal{H}$. Thus, if there exists $f \in M(\lambda, k, \gamma)$, $f \not\equiv 0$, then $\gamma = \pm 1$. At this point, this restriction upon γ is to be expected in light of the special case $a_n = b_n (n \in Z, n \geq 0)$ of

Theorem 2.1, since (as we have seen) the same restriction follows from the condition $\varphi = \psi$ in Definition 2.1

The parameter k is usually called *the weight of* f, and we adopt this standard terminology from the theory of automorphic/modular forms. Under further suitable growth restrictions, $f \in M(\lambda, k, \gamma)$ is called an *automorphic form of weight k with respect to the group $G(\lambda)$*. More specifically, $M_0(\lambda, k, \gamma)$ is the vector space of *entire* automorphic forms of weight k and multiplier γ with respect to $G(\lambda)$. See the beginning of Chapter 3 for the definition of $G(\lambda)$, and Chapter 5 for more about entire automorphic forms with respect to $G(\lambda)$ and for a discussion of the important subspace of "cusp forms." Since $G(1) = \Gamma(1) = SL(2, \mathbb{Z})$, the *modular group*, and $G(2)$ is a subgroup of index 3 in $\Gamma(1)$, in these cases f is often called *modular* rather than automorphic.

It is worth noting that for $\lambda < 2$, the assumption $k > 0$ is unnecessary in these definitions, since $k \leq 0$ implies that f is constant for $f \in M(\lambda, k, \gamma)$. (Actually, $f \equiv 0$, for $k < 0$ and for $k = 0$, $\gamma = -1$, as well.) The same holds true when $\lambda = 2$, if we assume that $f \in M_0(\lambda, k, \gamma)$. See Corollary 5.3 and Remark 6.3 for more details.

Our principal objective in Chapters 4–6 is to determine the dimension of the space $M_0(\lambda, k, \gamma)$. By Theorem 2.1, this problem is equivalent to finding the number of Dirichlet series with a given signature (λ, k, γ). If $\lambda > 2$, both $M_0(\lambda, k, \gamma)$ and $M(\lambda, k, \gamma)$ are infinite-dimensional. (See Theorem 4.1.) However, $M(\lambda, k, \gamma)$ is very much the larger of the two spaces. In fact, $M(\lambda, k, \gamma)/M_0(\lambda, k, \gamma)$ itself has infinite dimension (Theorem 4.2). On the other hand, if $\lambda < 2$, then $M_0(\lambda, k, \gamma) = M(\lambda, k, \gamma)$, and the space is of finite dimension, as we prove in Chapter 5. If $\lambda = 2$, then again $M(\lambda, k, \gamma)/M_0(\lambda, k, \gamma)$ has infinite dimension, but in contrast to the case $\lambda > 2$, when $\lambda = 2$ there is a natural condition, analogous to (i), which is equivalent to (iii) in the presence of (i) and (ii). See Definition 6.2 and Theorem 6.3, in which we formulate the condition and prove the equivalence, respectively.

In the sequel we restrict ourselves to the case $a_n = b_n$ (notation as in Theorem 2.1), in which event γ occurring in (2.1) is necessarily ± 1. If this restriction is lifted, one might ask whether a similar theory can be developed. For a discussion of this question, see Chapter 7.

Hamburger [39], [40] considered some functional equations very close in form to that satisfied by $\zeta(s)$. For example, replacing $\Gamma\left(\frac{1}{2}s\right)$ by $\Gamma\left(\frac{1}{2}(s+1)\right)$, he proved a result for Dirichlet L-functions analogous to the uniqueness theorem for $\zeta(s)$ discussed in our introduction. We might also ask about the uniqueness of Dirichlet series satisfying a functional equation with multiple gamma factors, such as the functional equation of a Dedekind zeta function (other than $\zeta(s)$ itself or the zeta function of an imaginary quadratic number field). The only result that we know is due to Maass [76, p. 145, Satz 2]. Under certain conditions he showed that the solution to a functional equation involving $\Gamma^2\left(\frac{1}{2}\{s+1\}\right)$ or $\Gamma^2\left(\frac{1}{2}s\right)$ is unique.

Chapter 3

A fundamental region

In the sequel we let

$$T = T(\tau) = -1/\tau$$

and

$$S_\lambda = S_\lambda(\tau) = \tau + \lambda,$$

where $\lambda > 0$.

Definition 3.1. The *Hecke group* $G(\lambda)$ is the group of linear fractional transformations generated by T and S_λ. We say that $f(\tau)$, $\tau \in \mathcal{H}$, is an *automorphic form on* $G(\lambda)$ if $f \in M(\lambda, k, \gamma)$. If $\lambda = 1$ or $\lambda = 2$, we may refer to $f(\tau)$ as a *modular form on* $G(\lambda)$.

With each transformation $V(\tau) = (a\tau + b)/(c\tau + d) \in G(\lambda)$ we associate the matrix

$$V = \begin{bmatrix} a & b \\ c & d \end{bmatrix}.$$

We identify V and $-V$. Thus, $G(\lambda)$ is isomorphic to the matrix group modulo $-I$, where I denotes the identity matrix.

Definition 3.2. We say that τ_1 and τ_2 are *equivalent with respect to* $G(\lambda)$ (or *modulo* $G(\lambda)$) if there exists a transformation $V \in G(\lambda)$ such that

$$V(\tau_1) = \tau_2.$$

It is easily checked that Definition 3.2 does define an equivalence relation.

Definition 3.3. A *fundamental region for* $G(\lambda)$ is an open subset R of $\mathcal{H}$ such that

(i) no two points of R are equivalent with respect to $G(\lambda)$;

(ii) every point in $\mathcal{H}$ is equivalent to some point in $\overline{R}$, the closure of R with respect to the extended complex plane.

Remark 3.1. It is an immediate consequence of the definition that if R is a fundamental region for $G(\lambda)$, then so is $M(R)$, for $M \in G(\lambda)$.

We now derive a fundamental region for $G(\lambda)$. The proof we give here is due to R. J. Evans [34].

Theorem 3.1. *Let $B(\lambda) = \{\tau \in \mathcal{H} : |x| < \lambda/2, |\tau| > 1\}$. Then if $\lambda \geq 2$ or if $\lambda = 2\cos(\pi/q)$, where $q \geq 3$ is an integer, $B(\lambda)$ is a fundamental region for $G(\lambda)$.*

Definition 3.4. Let $T_A = \{\lambda : \lambda = 2\cos(\pi/q), q \geq 3, q \in \mathbb{Z}\}$.

If $\lambda = 1$, i.e., $q = 3$, $G(1)$ is called *the modular group* in the literature. A derivation of the fundamental region in this case can also be found in several other texts, e.g., [60], [73], [74], or [37]. The advantage of the proof of Theorem 3.1 that we give here is that it is elementary, while the proofs in [48] and [87] appeal to advanced concepts in complex analysis.

We shall establish Theorem 3.1 through a series of lemmas.

Lemma 3.1. *Every point in $\mathcal{H}$ is equivalent with respect to $G(\lambda)$ to a point in $\overline{B(\lambda)}$.*

Proof. Define the following transformations on $\mathbb{C}$:

$$T_1(\tau) = \tau/|\tau|^2 = 1/\bar{\tau} \qquad \text{(reflection in the unit circle)},$$
$$T_2(\tau) = -\bar{\tau} \qquad \text{(reflection in the line } x = 0\text{)},$$

and

$$T_3(\tau) = -(\bar{\tau} + \lambda) \qquad \text{(reflection in the line } x = -\lambda/2\text{)}.$$

Note that $T_1T_2 = T$, $T_2T_3 = S_\lambda$, $T_1T_3 = -1/(\tau + \lambda)$, $T_j^2 = I$, $j = 1$, 2, 3, and $(T_iT_j)^{-1} = T_j^{-1}T_i^{-1} = T_jT_i$. Thus, it is easily seen that $G(\lambda)$ consists of all those words of even length involving T_1, T_2, and T_3. Hence, given an arbitrary point $\tau_0 \in \mathcal{H}$, it suffices to show that there exists a transformation $V \in \langle T_1, T_2, T_3 \rangle$, the group of transformations generated by T_1, T_2, and T_3, such that $V(\tau_0) \in \overline{B(\lambda)}$, for if $V \notin G(\lambda)$, then $T_2V \in G(\lambda)$ and $T_2V\tau_0 \in \overline{B(\lambda)}$.

Define a sequence of points $\{\tau_n\} = \{x_n + iy_n\}$ inductively as follows. Recalling that $T_2T_3 = S_\lambda$, apply T_2 and T_3, if necessary, to translate τ_0 to a point τ_1 in the strip $E(\lambda) = \{\tau \in \mathcal{H} : -\lambda/2 \leq x \leq 0\}$. Given $\tau_n (n \geq 1)$, apply T_2 and T_3 to translate $T_1\tau_n$ to a point $\tau_{n+1} \in E(\lambda)$. Assume that $|\tau_n| < 1$ for every n, for otherwise we are done. Let w be a cluster point of $\{\tau_n\}$. If $|w| < 1$, then $\{\tau_n\}$ has an infinite subsequence $\{\tau_{n_k}\}$ such that $|\tau_{n_k}| \leq c < 1$ for some positive number c, e.g., $c = (|w| + 1)/2$. Since $y_{n+1} = y_n/|\tau_n|^2$, $y_{n_k} \geq y_{n_1}/c^{2(k-1)}$, which tends to ∞ as k tends to ∞. Since this is a contradiction, we must have $|w| = 1$.

Since, with increasing n, τ_n approaches the unit circle, if $\lambda > 2$, $T_1\tau_n \in B(\lambda)$, for some n sufficiently large. The lemma is also clear for the case $\lambda = 2$. For since $\{y_n\}$ is increasing, $w \neq (-1, 0)$. Hence, again $T_1\tau_n \in B(2)$ for some n sufficiently large.

We now assume that $\lambda < 2$. Let τ_λ denote the intersection in $\mathcal{H}$ of the line $x = -\lambda/2$ and the unit circle $|\tau| = 1$. If $w \neq \tau_\lambda$, it is clear that, for some n sufficiently large, $T_1\tau_n \in B(\lambda)$. Thus, assume that $w = \tau_\lambda$. If for some n, $\arg(\tau_n) \leq \arg(\tau_\lambda)$, then since τ_λ and τ_n, $n \geq 1$, are in the left half-plane,

$$\operatorname{Im}(\tau_{n+1}) = \frac{\operatorname{Im}(\tau_n)}{|\tau_n|^2} \geq \frac{\sin(\arg\tau_\lambda)}{|\tau_n|} = \frac{\operatorname{Im}(\tau_\lambda)}{|\tau_n|} > \operatorname{Im}(\tau_\lambda).$$

However, this is a contradiction, as $\operatorname{Im}(\tau_n)$ is increasing monotonically to $\operatorname{Im}(\tau_\lambda)$. Hence, $\arg(\tau_n) > \arg(\tau_\lambda)$ for every n. Now, there exists a positive integer N such that for $n \geq N$, $\tau_{n+1} = T_3T_1\tau_n$, and so

$$x_{n+1} = -\lambda - x_n/|\tau_n|^2.$$

Note that $x_n \neq 0$, for otherwise $x_{n+1} = -\lambda \notin E(\lambda)$. Letting $\pi\theta = \pi - \arg(\tau_\lambda)$ (so that $\lambda = 2\cos\pi\theta$), we find that, for $n \geq N$,

$$\begin{aligned}
x_{n+1} - x_n &= -\frac{1}{x_n}\left(\lambda x_n + \frac{x_n^2}{x_n^2 + y_n^2} + x_n^2\right) \\
&= -\frac{1}{x_n}(\lambda x_n + \cos^2(\arg\tau_n) + x_n^2) \\
&> -\frac{1}{x_n}(\lambda x_n + \cos^2(\arg\tau_\lambda) + x_n^2) \\
&= -\frac{1}{x_n}(x_n + \cos\pi\theta)^2 \geq 0.
\end{aligned}$$

Thus, $x_{n+1} > x_n$ for every $n \geq N$. But this is a contradiction, as x_n tends to $\operatorname{Re}\tau_\lambda$ and $x_n \geq \operatorname{Re}\tau_\lambda$. □

We must now show that for the values of λ given in Theorem 3.1 no two points of $B(\lambda)$ are equivalent.

Lemma 3.2. *For $\lambda \geq 2$, no two distinct points of $B(\lambda)$ are equivalent.*

Proof. Let $V \in G(\lambda)$, where $V \neq I$. We can write V in the form

$$V = S_\lambda^{k_r} T S_\lambda^{k_{r-1}} \cdots S_\lambda^{k_2} T S_\lambda^{k_1},$$

where $r \geq 1$, $k_i \in \mathbb{Z}$, $1 \leq i \leq r$, and $k_i \neq 0$, $2 \leq i \leq r-1$. Let τ_0 denote an arbitrary point in $B(\lambda)$ and define for $1 \leq i \leq r-1$,

$$\tau_i = T S_\lambda^{k_i} T S_\lambda^{k_{i-1}} \cdots T S_\lambda^{k_1} \tau_0.$$

It is easily seen that $|\tau_i| < 1$ for $1 \leq i \leq r-1$. Thus, $V\tau_0 = S_\lambda^{k_r} \tau_{r-1} \notin B(\lambda)$, and the proof is complete. □

We must still examine the case when $\lambda \in T_A$; see Definition 3.4.

Lemma 3.3. *If $\lambda \in T_A$, no two points of $B(\lambda)$ are equivalent under a nonidentity transformation in $\langle T_1, T_3 \rangle$.*

Proof. If the lemma is false, then there exist points τ, $\tau' \in B(\lambda)$ and a transformation $V \in \langle T_1, T_3 \rangle$ such that $V\tau = \tau'$. Note that $V \neq T_3$, for otherwise $T_3\tau \notin B(\lambda)$. Write, as before, $\lambda = 2\cos \pi\theta$, where for the moment $\lambda < 2$ is arbitrary. An easy exercise in induction on n shows that

$$(T_1, T_3)^n = \begin{bmatrix} a_n & b_n \\ c_n & d_n \end{bmatrix} = \frac{1}{\sin \pi\theta} \begin{bmatrix} \sin \pi\theta(1-n) & -\sin \pi\theta n \\ \sin \pi\theta n & \sin \pi\theta(n+1) \end{bmatrix}. \tag{3.1}$$

If $\lambda \in T_A$, then $\theta = 1/q$, and if $n = q$, we see that $(T_1T_3)^q = I$, the identity transformation. Hence, we can write V in the form

$$V = T_3^\alpha (T_1T_3)^n,$$

where $\alpha = 0$ or 1 and $n \in \mathbb{Z}$ with $1 \leq |n| \leq q-1$. (Note that if $V = (T_1T_3)^n T_3$, then $V = T_3(T_1T_3)^{-n}$.) From (3.1) we see that for $\theta = 1/q$, $c_n d_n \geq 0$. Therefore,

$$\begin{aligned} |c_n\tau + d_n|^2 &= c_n^2|\tau|^2 + d_n^2 + 2c_nd_nx \\ &> c_n^2 + d_n^2 - \lambda c_nd_n = 1, \end{aligned}$$

upon a tedious, but elementary, calculation. Thus,

$$\begin{aligned} \operatorname{Im}(\tau') &= \operatorname{Im}(T_3^\alpha (T_1T_3)^n \tau) \\ &= \operatorname{Im}\{(T_1T_3)^n \tau\} = \frac{y}{|c_n\tau + d_n|^2} < y = \operatorname{Im}(\tau). \end{aligned}$$

In summary, we have shown that if τ and τ' are two points in $B(\lambda)$ such that $V\tau = \tau'$ for some $V \in G(\lambda)$, $V \neq I$, then $\mathrm{Im}(\tau') = \mathrm{Im}(V\tau) < \mathrm{Im}(\tau)$. Now repeat the same argument with the roles of τ and τ' reversed and V replaced by V^{-1}. We conclude that $\mathrm{Im}(\tau) = \mathrm{Im}(V^{-1}\tau') < \mathrm{Im}(\tau')$. Hence, we have a contradiction, and the lemma is proved. □

Lemma 3.4. *Let $\lambda \in T_A$, $\tau \in \mathcal{H}$, and $W \in \langle T_1, T_3\rangle$, where $W \neq I$ and $W \neq T_1$. If either*

(i) $\mathrm{Re}\,\tau > 0$

or

(ii) $\tau \in B(\lambda)$,

then $\mathrm{Re}(W\tau) < 0$.

Proof. As before, since $(T_1T_3)^q = I$, we can write W in the form

$$W = T_1^{\alpha}(T_1T_3)^n,$$

where $\alpha = 0$ or 1 and $n \in \mathbb{Z}$ with $1 \leq |n| \leq q-1$. To show that $\mathrm{Re}(W\tau) < 0$, it suffices to show that $\mathrm{Re}(T_1T_3)^n\tau < 0$. Now, using the notation of (3.1), we see that

$$\mathrm{Re}(T_1T_3)^n\tau = \frac{(a_nx + b_n)(c_nx + d_n) + a_nc_ny^2}{|c_n\tau + d_n|^2}. \tag{3.2}$$

From (3.1) we also note that a_n and b_n share a sign opposite to that shared by c_n and d_n. Hence, if (i) holds, $a_nc_ny^2 \leq 0$ and $(a_nx+b_n)(c_nx+d_n) < 0$, and so from (3.2) we see that $\mathrm{Re}(T_1T_3)^n\tau < 0$. If (ii) holds, since $a_nc_n \leq 0$ and $(a_nd_n + b_nc_n) \leq 0$,

$$\begin{aligned}\mathrm{Re}(T_1T_3)^n\tau &= \frac{a_nc_n|\tau|^2 + b_nd_n + (a_nd_n + b_nc_n)x}{|c_n\tau + d_n|^2} \\ &\leq \frac{a_nc_n + b_nd_n + (a_nd_n + b_nc_n)(-\lambda/2)}{|c_n\tau + d_n|^2} \\ &= -\frac{\cos(\pi/q)}{|c_n\tau + d_n|^2} < 0,\end{aligned}$$

by a straightforward calculation with the use of (3.1). This establishes the lemma. □

Lemma 3.5. *If $\lambda \in T_A$, no two distinct points of $B(\lambda)$ are equivalent.*

Proof. It is sufficient to show that no two points are equivalent under a transformation $V \in \langle T_1, T_2, T_3\rangle$, where $V \neq I$ and $V \neq T_2$. Suppose that the contrary is true. We shall then choose a transformation $V \in \langle T_1, T_2, T_3 \rangle$, $V \neq T_2$, I, of minimal length for which there exists a point $\tau \in B(\lambda)$ such that $V\tau \in B(\lambda)$. By Lemma 3.3, the word V must contain T_2. No representation of V can begin or end with T_2. For if $V = T_2 Y$, then $Y \neq T_2$, $Y \neq I$, and $Y\tau \in B(\lambda)$, which contradicts the minimality of V. Similarly, if $V = YT_2$, then $Y \neq T_2$, $Y \neq I$, and $Y(T_2\tau) \in B(\lambda)$, which again contradicts the minimality of V, since $T_2\tau \in B(\lambda)$. Since $T_1T_2 = T_2T_1$, we replace all pairs T_1T_2 occurring in V by T_2T_1. Note that V cannot end with T_2T_1, for if $V = YT_2T_1$, then V would have the representation $V = YT_1T_2$, and we have just shown that V cannot end in T_2. Thus, V must have the form,

$$V = W_1 T_2 W_2 T_2 \cdots W_k T_2 W_{k+1},$$

where $W_i \in \langle T_1, T_3 \rangle$, $W_i \neq I$, and $W_i \neq T_1$, $1 \leq i \leq k+1$. For $2 \leq i \leq k+1$, let

$$\tau_i = T_2 W_i T_2 W_{i+1} \cdots W_k T_2 W_{k+1} \tau. \tag{3.3}$$

We shall show by induction on i that $\text{Re}(\tau_i) < 0$, for $2 \leq i \leq k+1$. Since $V\tau \in B(\lambda)$, $\text{Re}(\tau_2) = \text{Re}(W_1^{-1}V\tau) < 0$ by (3.3) and Lemma 3.4. Assume that $\text{Re}(\tau_m) < 0$ for some m, $2 \leq m \leq k$. Then $\text{Re}(T_2\tau_m) > 0$. By (3.3) and Lemma 3.4, $\text{Re}(\tau_{m+1}) = \text{Re}(W_m^{-1}T_2\tau_m) < 0$, and the induction is complete. Since $\tau \in B(\lambda)$, $\text{Re}(W_{k+1}\tau) < 0$ by Lemma 3.4. Thus, from (3.3), $\text{Re}(\tau_{k+1}) = \text{Re}(T_2 W_{k+1}\tau) > 0$. But we previously showed that $\text{Re}(\tau_{k+1}) < 0$, and so we have a contradiction. This completes the proof of Lemma 3.5 and thus of Theorem 3.1 as well. □

Remark 3.2. We show that if $0 < \lambda < 2$, $\lambda \notin T_A$, then there exist two points in $B(\lambda)$ that are equivalent under a transformation in $G(\lambda)$. Thus, $B(\lambda)$ cannot be a fundamental region for $G(\lambda)$. Actually, much more is true: if $0 < \lambda < 2$, but $\lambda \notin T_A$, then $G(\lambda)$ is not discontinuous and so not discrete. For a discussion of discontinuous and discrete groups, see [73, Chapter 3] and [74, Chapter 1]. See also the discussion in Chapter 5, immediately before the statement of Theorem 5.5, especially Theorem 5.3.

Because $\lambda = 2\cos\pi\theta \notin T_A$, we can find an integer k such that $1/(k+1) < \theta < 1/k$. Return again to (3.1). Letting $\tau = iy$, $y > 1$, we find that

$$\tau' = (T_1T_3)^k \tau = \frac{b_k d_k + a_k c_k y^2}{c_k^2 y^2 + d_k^2} + i \frac{y}{c_k^2 y^2 + d_k^2}.$$

From a calculation in the proof of Lemma 3.3, $c_k^2 + d_k^2 - \lambda c_k d_k = 1$. By our choice of k, c_k and d_k have opposite signs. Thus, $c_k^2 + d_k^2 = 1 + \lambda c_k d_k < 1$. Also, choose y such that $\mathrm{Re}(\tau')$ is not an integral multiple of $\lambda/2$. Then, $\mathrm{Im}(\tau') > y > 1$, and an appropriate power of S_λ translates τ' to a point $\tau'' \in B(\lambda)$. Thus, τ and τ'' are the desired equivalent points in $B(\lambda)$.

Chapter 4

The case $\lambda > 2$

Theorem 4.1. *If* $\lambda > 2$, *then* $\dim M_0(\lambda, k, \gamma) = \infty$ *for every* $k > 0$ *and* $\gamma = \pm 1$.

Remark 4.1. Intuitively, the reason $\dim M_0(\lambda, k, \gamma) = \infty$ is as follows. The conditions in Definition 2.2 that f must satisfy are restrictions on f in the upper half-plane $\mathcal{H}$. For $\lambda > 2$, the fundamental region for $G(\lambda)$ in Theorem 3.1 is bounded by the line segments $(-\lambda/2, -1)$ and $(1, \lambda/2)$ on the real axis. This will enable us to continue $f \in M_0(\lambda, k, \gamma)$ analytically from a fundamental region into the lower half-plane where singularities may be given to f. For $\lambda \leq 2$, a fundamental region is not bounded by any line segment or segments on the real axis. Thus, such a continuation is impossible for $\lambda \leq 2$, and, in fact, the real axis is a natural boundary for f.

Proof of Theorem 4.1. Let

$$B^*(\lambda) = \{\tau : -\lambda/2 < x < 0, |\tau| > 1\}.$$

Since $B^*(\lambda)$ is simply connected and since the open unit disc and $\mathcal{H}$ are conformally equivalent, by the Riemann mapping theorem there exists a function $z = g(\tau)$ that maps $B^*(\lambda)$ one-to-one and conformally onto $\mathcal{H}$. Now, a mapping from the open unit disc onto $\mathcal{H}$ can be effected by a linear fractional transformation. This linear fractional transformation can be uniquely determined by specifying three values on the boundary. Specifying the values of three points on the boundary of $B^*(\lambda)$ uniquely determines a conformal mapping h from $B^*(\lambda)$ onto the open unit disc. This can be seen from mapping the disc onto $B^*(\lambda)$ and then from $B^*(\lambda)$ onto the disc again. For if h is not uniquely determined, then we can find in the manner just described a nonidentity conformal map from the disc onto the disc that fixes three boundary points. But this is impossible. Thus, $g(\tau)$ may be uniquely chosen so that

$$g(i\infty) = \lim_{y\to\infty} g(x+iy) = 1,$$
$$g(i) = 0,$$

and

$$g(-i) = \infty.$$

The function g maps the boundary of $B^*(\lambda)$ onto the real axis. In fact, g takes the line segment $[i, i\infty)$ onto $[0,1)$, the left half of the unit circle onto the negative real axis, the line segment $(-i\infty, -i)$ onto (a_0, ∞) for some a_0, $1 < a_0 < \infty$, and the line $x = -\lambda/2$ onto $[1, a_0]$.

We now continue g analytically into the entire complex plane except for a set of points on the real axis. We define

$$g(T_i\tau) = \overline{g(\tau)}, \quad i = 1, 2, 3, \tag{4.1}$$

and by the Schwarz reflection principle obtain an analytic continuation of g onto $T_iB^*(\lambda)$, $i = 1, 2, 3$. (Recall that T_1, T_2, and T_3 are the three reflections defined at the beginning of the proof of Lemma 3.1.) The continuation to the whole complex plane, except for a set of points on the real axis, is obtained by iteration. The continuation is well defined because $B(\lambda)$ is a fundamental region for $G(\lambda)$. In particular, if n is even, then

$$g\left(\prod_{k=1}^{n} T_{i_k}\tau\right) = g(\tau), \tag{4.2}$$

where $i_k = 1$, 2, or 3. Since $G(\lambda)$ consists of all words of even length in T_1, T_2, and T_3, g is then invariant under transformations of $G(\lambda)$. We note that the analyticity of g in $\mathcal{H}$ follows directly from the definition of g and the Schwarz reflection principle. Since g is analytic on $(B^*(\lambda)) \cap (\mathcal{H} \cup \mathbb{R})$, and since $g(i\infty) = 1$, g is bounded on $B^*(\lambda) \cap \mathcal{H}$. Hence, g is bounded on $\mathcal{H}$.

Remark 4.2. On the entire Riemann sphere, g is analytic except at:
(a) points equivalent to $-i$ with respect to $G(\lambda)$;
(b) exceptional points on $\mathbb{R}$ mentioned above (4.1);
(c) limit points of the set of all points described in (a) and (b).

Note that the symbol $i\infty$, as distinct from ∞, designates the point ∞ approached only in the positive vertical direction. The symbol $-i\infty$ is analogous.

We now continue the proof of Theorem 4.1 with a careful examination of g at the "corners" $\pm i$ and $\pm i\infty$ of $B^*(\lambda)$. The transformation $T\tau = -1/\tau$ has the two fixed points $\pm i$. The transformation $S_\lambda \tau = \tau + \lambda$ has $\pm i\infty$ as its only fixed points. It then follows from (4.1) and (4.2) that g is one-to-one in a neighborhood of every point except those equivalent to $\pm i$ or $\pm i\infty$. Before proceeding further we give the following definition.

Definition 4.1. Suppose that $\tau_0 \in \mathbb{C} \cup \{i\infty\}$ and $\lambda > 0$. A function $t = t(\tau)$ is called *a local uniformizing variable at* τ_0 *with respect to* $G(\lambda)$ if $t(\tau)$ is analytic in a punctured neighborhood of τ_0 and $t(\tau)$ has the additional property:

> There exists a subset $\mathcal{N}(\tau_0, \lambda) = \mathcal{N}$ of some punctured neighborhood U' of τ_0 such that (i) $t(\tau)$ maps $\mathcal{N}$ one-to-one and onto a punctured disc D', $0 < |t| < \epsilon$, with $\epsilon > 0$; (ii) distinct points of $\mathcal{N}$ are inequivalent modulo $G(\lambda)$; and (iii) $\mathcal{N}$ is maximal in U' with respect to property (ii).

We note that $\pm i$ are fixed points of the group $G(\lambda)$ since $T \in G(\lambda)$; consequently,

$$V(i), V(-i) = \overline{V(i)}, \qquad \forall\, V \in G(\lambda), \tag{4.3}$$

are fixed points as well. If $\mathcal{H}^-$ denotes the lower half-plane, and if $\tau_0 \in \mathcal{H} \cup \mathcal{H}^-$ is *not* of the form (4.3), it follows directly from the shape of the fundamental region $B(\lambda)$ (see Theorem 3.1) and the fact that, for any M in $G(\lambda)$, $M(B(\lambda))$ is again a fundamental region of $G(\lambda)$, that there exists a full neighborhood $\mathcal{N}$ of τ_0 such that $V(\mathcal{N}) \cap \mathcal{N} = \phi$ for all $V \in G(\lambda)$, $V \neq I$. We conclude that:

(a) τ_0 is not a fixed point of $V \in G(\lambda)$, other than $V = I$;

(b) $t(\tau) = \tau - \tau_0$ serves as a local uniformizing variable at τ_0, since distinct points of $\mathcal{N}$ are inequivalent modulo $G(\lambda)$.

Remark 4.3. By way of contrast, if $\tau_0 \in \mathcal{H} \cup \mathcal{H}^-$ *is* of the form (4.3), then $t(\tau) = \tau - \tau_0$ is *not* a local uniformizing valuable at τ_0, since in this case any neighborhood of τ_0 necessarily contains distinct points equivalent with respect to $G(\lambda)$. The appropriate choice at τ_0 of the form (4.3) is, in fact, $t = \left(\frac{\tau - \tau_0}{\tau - \bar{\tau}_0}\right)^2$. (See the discussion below for the case $\tau_0 = i$.)

We return now to the proof of Theorem 4.1 and examine g in a neighborhood of i. Let

$$t_1(\tau) = \frac{\tau - i}{\tau + i}.$$

Then,

$$t_1(-1/\tau) = \frac{-1/\tau - i}{-1/\tau + i} = -\frac{i(\tau - i)}{i(\tau + i)} = -t_1(\tau). \tag{4.4}$$

We claim that $t(\tau) = t_1(\tau)^2$ is a local uniformizing variable for $G(\lambda)$ at $\tau_0 = i$. To verify this, let $\rho \in \mathbb{R}$, $0 < \rho < 1$, and let U be the open disc with diameter the line segment joining $i(1+\rho)$ to $i(1+\rho)^{-1}$ on the imaginary axis. Then $t_1(\tau)$ maps U one-to-one onto the disc $D : |t_1| < \frac{\rho}{2+\rho}$, with the portion U^+ of U above the unit circle mapping to the right half of D and the portion U^- below the unit circle mapping to the left half of D. (Observe that the center of U is $\frac{i}{2}\{(1+\rho) + (1+\rho)^{-1}\}$, a point above i on the imaginary axis, and $t_1(\tau)$ does not map the center of U to 0, the center of D.)

Since $\lambda > 2$ and $\rho < 1$, it follows easily that $U^+ \subset B(\lambda)$, so that $U^- \subset T(B(\lambda))$, the latter being again a fundamental region for $G(\lambda)$. Thus, defining $\mathcal{N} = \mathcal{N}(i, \lambda)$ to be U^+ adjoined to the open arc of the unit circle from i to the point of intersection of the unit circle with the right side of U, we find that $t(\tau) = t_1(\tau)^2 = \left(\frac{\tau - i}{\tau + i}\right)^2$ satisfies the Definition 4.1 of "local uniformizing variable at i."

Now observe that $g(\tau)$ can be rewritten as a function of t_1, since the mapping $\tau \to t_1$ is one-to-one on the Riemann sphere (thus, one-to-one in a neighborhood of i); let $g(\tau) = \hat{g}(t_1)$, say. On the other hand, by (4.4), the invariance property $g(-1/\tau) = g(\tau)$ is equivalent to: $\hat{g}(t_1)$ is an even function of t_1. Thus, in fact, $g(\tau) = \hat{g}(t_1) = g_1(t_1^2) = g_1(t)$. If we write

$$g_1(t) = \sum_{n=0}^{\infty} a_n t^n,$$

then $a_0 = 0$, since $g(i) = 0$, and $a_1 \neq 0$, as $g_1(t)$ is one-to-one in a neighborhood of $t = 0$. We conclude that $g_1(t)$ has a simple zero at $t = 0$, i.e., $g(\tau)$ has a double zero at $\tau = i$.

Remark 4.4. Note that while $\left(\frac{\tau - i}{\tau + i}\right)^4$ satisfies conditions (i) and (ii) of the definition of a local variable at i with respect to $G(\lambda)$, it fails to satisfy the maximality condition (iii). The latter is designed to ensure that the invariant function $g(\tau)$ can, in fact, be expressed in terms of t near $t = 0$, that is, near $\tau = i$. We have observed above that the invariance property $g(-1/\tau) = g(\tau)$ is equivalent to the expressibility of $g(\tau)$ as a function of $\left(\frac{\tau-i}{\tau+i}\right)^2$. However, there is no basis for concluding that $g(\tau)$ can be written as a function of $\left(\frac{\tau-i}{\tau+i}\right)^4$.

To examine g in a neighborhood of $\tau = -i$, put

$$t_2(\tau) = \frac{\tau + i}{\tau - i}.$$

Proceeding as above, we find that $t = t_2^2$ is a local uniformizing variable for g in a neighborhood of $\tau = -i$ with respect to $G(\lambda)$. Since $g(\tau)$ is not analytic at $\tau = -i$, $g(\tau) = g_2(t)$, say, is not analytic at $t = 0$. But since $g_2(t)$ is one-to-one near $t = 0$, g_2 must have a simple pole at $t = 0$, i.e., $g(\tau)$ has a double pole at $\tau = -i$.

We turn to the behavior of $g(\tau)$ in a punctured neighborhood of $i\infty$, specifically in an open half-plane of the form $\operatorname{Im}\tau > B$, $B > 0$. Choose $B > 1$, say, for convenience. With

$$\mathcal{N} = \mathcal{N}(i\infty, \lambda) := \{\tau : \operatorname{Im}\tau > B, -\lambda/2 \leq \operatorname{Re}\tau < \lambda/2\},$$

we see that

$$t(\tau) = e^{2\pi i\tau/\lambda}$$

satisfies the conditions (i)–(iii) for a local uniformizing variable at $i\infty$ with respect to $G(\lambda)$. As one might expect, the periodicity of g, i.e., $g(\tau + \lambda) = g(\tau)$, guarantees that

$$g(\tau) = g_3(t) = \sum_{n=0}^{\infty} a_n t^n, \tag{4.5}$$

in a neighborhood of $t = 0$.

To verify (4.5), note that, ignoring the branching of the logarithm, we have $\tau = \frac{\lambda}{2\pi i}\log t$. This implies that $g(\tau) = g\left(\frac{\lambda}{2\pi i}\log t\right)$, a (possibly multiple-valued) function of t. Since the distinct branches of $\log t$ differ by $2\pi i n$, $n \in \mathbb{Z}$, it suffices to consider only

$$g\left(\frac{\lambda}{2\pi i}(\log t + 2\pi i n)\right) = g\left(\frac{\lambda}{2\pi i}\log t + n\lambda\right),$$

together with $g\left(\frac{\lambda}{2\pi i}\log t\right)$. But the periodicity of g shows that all of these branches of $g\left(\frac{\lambda}{2\pi i}\log t\right)$ are equal, so that g is a single-valued function of t holomorphic in a punctured neighborhood of $t = 0$. Since g is bounded near $t = 0$ as well, (4.5) follows. In addition, $a_0 = 1$, because $g(i\infty) = 1$, and $a_1 \neq 0$, since g is one-to-one on $B^*(\lambda)$.

We now claim that to prove Theorem 4.1 it suffices to show that $\dim M_0(\lambda, k, \gamma) \neq 0$. For if $f \in M_0(\lambda, k, \gamma)$, the same is true of fg^n, $n = 0$, 1, 2, ..., provided that we can show that the requirements of Definition 2.2 are satisfied by fg^n. First, (i) is clear from (4.5). Secondly, (ii) is clear from the invariance property $g(-1/\tau) = g(\tau)$. We have already seen that $g(\tau)$ is bounded on $\mathcal{H}$. Hence, $g(\tau) = O(1)$ as y tends to 0, uniformly for $\tau \in \mathcal{H}$. From Lemma 2.2, (iii) of Definition 2.2 is seen to be satisfied by fg^n. Thus, $fg^n \in M_0(\lambda, k, \gamma)$, $n = 0, 1, 2, \ldots$. Furthermore, these functions are linearly independent since g^n has a pole of order $2n$ at $\tau = -i$.

Now, in a neighborhood of $\tau = i$,

$$\begin{aligned} g(\tau) &= c_1(\tau - i)^2 + \cdots, \qquad c_1 \neq 0, \\ &= (\tau - i)^2 g^*(\tau), \end{aligned}$$

where $g^*(\tau)$ is analytic in a neighborhood of $\tau = i$ and $g^*(i) \neq 0$. Thus, we can define a single-valued, analytic square root in a neighborhood of $\tau = i$. Of course, we can do the same at all points equivalent to i. Since g is analytic and nonzero on the remainder of $\mathcal{H}$, an analytic square root may be defined in a neighborhood of each of these other points as well. Since $\mathcal{H}$ is simply connected and $g(\tau)^{\frac{1}{2}}$ is locally single-valued (hence locally analytic) at each point of $\mathcal{H}$, the monodromy theorem permits us to define a single-valued square root $G(\tau) = \{g(\tau)\}^{\frac{1}{2}}$ on all of $\mathcal{H}$.

As g has period λ, it follows that $G(\tau + \lambda) = \pm G(\tau)$. We claim that, in fact, $G(\tau + \lambda) = G(\tau)$. To prove this, assume, by way of contradiction, that $G(\tau + \lambda) = -G(\tau)$, and define

$$h(\tau) = e^{-\pi i\tau/\lambda} G(\tau).$$

Then, $h(\tau + \lambda) = h(\tau)$, and the argument used to derive (4.5) yields

$$h(\tau) = \sum_{n=-\infty}^{\infty} b_n e^{2\pi i n\tau/\lambda},$$

that is,

$$G(\tau) = \sum_{n=-\infty}^{\infty} b_n e^{\pi i(2n+1)\tau/\lambda}. \tag{4.6}$$

Since $g(i\infty) = 1$, we deduce that $G(i\infty) = \pm 1$. In any event, G is bounded at $i\infty$, so (4.6) reduces to

$$G(\tau) = \sum_{n=0}^{\infty} b_n e^{\pi i(2n+1)\tau/\lambda},$$

and we conclude that $G(i\infty) = 0$. This contradiction implies that $G(\tau + \lambda) = G(\tau)$.

Since $g(\tau) = g(-1/\tau)$, we find that

$$G(\tau) = \pm G(-1/\tau). \tag{4.7}$$

We must determine which sign is correct. In a neighborhood of $\tau = i$, we know from our discussion above that

$$G(\tau) = t_1(\tau) G_1(\tau),$$

where $G_1(\tau)$ is analytic and $G_1(i) \neq 0$. Thus, from (4.4),

$$\frac{G(\tau)}{G(-1/\tau)} = \frac{t_1(\tau)G_1(\tau)}{t_1(-1/\tau)G_1(-1/\tau)} = -\frac{G_1(\tau)}{G_1(-1/\tau)}.$$

Letting τ tend to i above, we conclude that the minus sign in (4.7) is correct, i.e.,

$$G(\tau) = -G(-1/\tau). \tag{4.8}$$

Now define

$$H(\tau) = \frac{g'(\tau)}{G(\tau)\{g(\tau) - 1\}}.$$

Since g, g', and G have period λ, then

$$H(\tau) = H(\tau + \lambda). \tag{4.9}$$

From (4.8) and the fact that

$$g'(\tau) = g'(-1/\tau)/\tau^2,$$

we deduce that

$$H(-1/\tau) = \frac{\tau^2 g'(\tau)}{-G(\tau)\{g(\tau)-1\}} = (\tau/i)^2 H(\tau). \tag{4.10}$$

Since g and G are analytic on $\mathcal{H}$, $H(\tau)$ is analytic on $\mathcal{H}$, save for possibly those points where $g(\tau) = 1$ or $g(\tau) = 0$. Since $g(i\infty) = 1$ and $g(\tau)$ is one-to-one on $B^*(\lambda)$, there are no points in $\mathcal{H}$ where $g(\tau) = 1$. If

$$g(\tau) = 1 + b_1 e^{2\pi i\tau/\lambda} + \cdots, \qquad b_1 \neq 0,$$

then

$$g'(\tau) = \frac{b_1 2\pi i}{\lambda} e^{2\pi i\tau/\lambda} + \cdots.$$

Hence, since $g(\tau)-1$ has a simple zero at $i\infty$, $H(i\infty)$ exists and $H(i\infty) \neq 0$. Since g has a double zero at $\tau = i$, G and g' have simple zeros at $\tau = i$. Hence, $H(\tau)$ is analytic at $\tau = i$ and $H(i) \neq 0$. We conclude that $H(\tau)$ is analytic on $\mathcal{H}$ and has no zeros on $\mathcal{H} \cup \{i\infty\}$.

Now, (4.9), (4.10), and the fact that $H(i\infty)$ exists show that $H(\tau) \in M(\lambda, 2, 1)$. Also, since $H(\tau) \neq 0$ on $\mathcal{H}$ and since $\mathcal{H}$ is simply connected, we can define an analytic function

$$f_k(\tau) = \{H(\tau)\}^{k/2}$$

on $\mathcal{H}$. Since $H(\tau)$ has period λ, clearly we see that for some constant ϵ_1, $|\epsilon_1| = 1$,

$$f_k(\tau + \lambda) = \epsilon_1 f_k(\tau). \tag{4.11}$$

It is also clear from (4.10) that

$$f_k(-1/\tau) = \epsilon_2 (\tau/i)^k f_k(\tau), \tag{4.12}$$

for some constant ϵ_2, $|\epsilon_2| = 1$. We now determine ϵ_1 and ϵ_2. In (4.11) let τ approach $i\infty$. Since $f_k(\tau)$ and $f_k(\tau + \lambda)$ approach the same nonzero value, we conclude that $\epsilon_1 = 1$. Since $H(i) \neq 0$, we can determine ϵ_2 by letting $\tau = i$ in (4.12). Hence, $\epsilon_2 = 1$. We have therefore shown that $f_k \in M(\lambda, k, 1)$.

Consider

$$f(\tau; n) = f_k(\tau)\{g(\tau) - 1\}^n, \tag{4.13}$$

where $n \in \mathbb{Z}$, $n > k/2$. We claim that, for $n > k/2$, $f(\tau; n) \in M_0(\lambda, k, 1)$. Since $g(\tau) = g(-1/\tau)$ and g is bounded on $\mathcal{H}$, we see that $\{g(\tau) - 1\}^n$ lies in $M_0(\lambda, 0, 1)$. Now since $f_k \in M(\lambda, k, 1)$, it follows that $f(\tau; n) \in M(\lambda, k, 1)$.

To prove that $f(\tau; n) \in M_0(\lambda, k, 1)$ for $n > k/2$, examine

$$f^*(\tau) = |\tau - \bar{\tau}|\ |g'(\tau)/G(\tau)| = 2y|g'(\tau)/G(\tau)|.$$

Let $V \in G(\lambda)$, where $V\tau = (a\tau + b)/(c\tau + d)$. Since $g(V\tau) = g(\tau)$, we find that

$$\begin{aligned} g'(\tau) = \frac{dg(V\tau)}{d\tau} &= g'(V\tau)\left\{\frac{a}{c\tau + d} - \frac{(a\tau + b)c}{(c\tau + d)^2}\right\} \\ &= g'(V\tau)\left\{\frac{ac\tau + ad - ac\tau - bc}{(c\tau + d)^2}\right\} \\ &= \frac{g'(V\tau)}{(c\tau + d)^2}. \end{aligned}$$

Thus,

$$\begin{aligned} f^*(V\tau) &= |2\operatorname{Im}(V\tau)|\ |g'(V\tau)/G(V\tau)| \\ &= \frac{2y}{|c\tau + d|^2} \cdot \frac{|g'(\tau)|\ |c\tau + d|^2}{|G(\tau)|} \\ &= 2y|g'(\tau)/G(\tau)| \\ &= f^*(\tau). \end{aligned} \tag{4.14}$$

Clearly, $yg'(\tau)$ tends to 0 as τ tends to $i\infty$. Also, since g' is analytic on $B^*(\lambda) \cap (\mathcal{H} \cup R)$, it follows that $g'(\tau)$ is bounded on $B^*(\lambda) \cap \mathcal{H}$. Furthermore, the only zero of $G(\tau)$, a simple one at $\tau = i$, is cancelled by the simple zero of $g'(\tau)$ at $\tau = i$. Hence, $f^*(\tau)$ is bounded on $B^*(\lambda) \cap \mathcal{H}$. By (4.14), we conclude that $f^*(\tau)$ is bounded on $\mathcal{H}$.

As $n > k/2$, $\{g(\tau) - 1\}^{n-k/2}$ is bounded on $\mathcal{H}$ since g is bounded on $\mathcal{H}$. Using also the boundedness of f^*, we conclude that

$$|f(\tau; n)(2y)^{k/2}| = |f^*(\tau)|^{k/2}|g(\tau) - 1|^{n-k/2} = O(1),$$

or

$$f(\tau; n) = O(y^{-k/2}),$$

uniformly in x as $y > 0$ tends to 0. Hence, by Lemma 2.2, $f(\tau; n) \in M_0(\lambda, k, 1)$. Since $f(\tau, n) \not\equiv 0$, it follows from our earlier observation that $M_0(\lambda, k, 1)$ has infinite dimension.

It is now easy to show that $\dim M_0(\lambda, k, -1) = \infty$. Consider

$$g_k(\tau) = f_k(\tau)G(\tau).$$

Now from (4.8) and (4.12),

$$\begin{aligned} g_k(-1/\tau) = f_k(-1/\tau)G(-1/\tau) &= -(\tau/i)^k f_k(\tau)G(\tau) \\ &= -(\tau/i)^k g_k(\tau). \end{aligned}$$

Clearly, $g_k(\tau+\lambda) = g_k(\tau)$. Now proceed in the same manner with $g_k(\tau)$ as we did with $f_k(\tau)$. We conclude that $g_k \in M_0(\lambda, k, -1)$ and that $\dim M_0(\lambda, k, -1) = \infty$. The proof of Theorem 4.1 is now complete. □

Remark 4.5. A careful examination of the proof of Theorem 4.1 shows that it does not depend upon the assumption $k > 0$ in an essential way, so that this result holds for all real k, with $\gamma = \pm 1$, as long as $\lambda > 2$. For $k \leq 0$ we need to observe that $1/f^*(\tau)$ (in place of $f^*(\tau)$) is bounded in $\mathcal{H}$. We then simply replace the function $f_k(\tau) = \{H(\tau)\}^{k/2}$ by $\hat{f}_k(\tau) = \{H(\tau)\}^{-k/2}$, with $H(\tau)$ as before, to conclude that, in this case, $\hat{f}(\tau; n) = \hat{f}_k(\tau)\{g(\tau) - 1\}^n = O(y^{k/2})$. The remainder of the argument proceeds as before. The situation in Theorem 4.2 is a bit different, as the proof below is completely independent of the sign of k.

As a corollary to the proof of Theorem 4.1, we establish the following theorem.

Theorem 4.2. *For $\lambda > 2$, k real, and $\gamma = \pm 1$,*

$$\dim M(\lambda, k, \gamma)/M_0(\lambda, k, \gamma) = \infty.$$

Remark 4.6. Since $M(\lambda, k, \gamma) \supset M_0(\lambda, k, \gamma)$, Theorem 4.1 implies that $\dim M(\lambda, k, \gamma) = \infty$. Of course, Theorem 4.2 is an even stronger statement concerning the size of $M(\lambda, k, \gamma)$.

Proof of Theorem 4.2. We are required to show that there exists an infinite collection of functions $\varphi_n(\tau)$ in $M(\lambda, k, \gamma)$ with the property:

$$\begin{aligned} &\text{if } c_1\varphi_{n_1} + \cdots + c_p\varphi_{n_p} \in M_0(\lambda, k, \gamma), \\ &\text{with distinct } n_j, \text{ then } c_1 = \cdots = c_p = 0. \end{aligned} \tag{4.15}$$

We begin by defining a modified version $\tilde{g}(\tau)$ of the mapping function $g(\tau)$ occurring in Theorem 4.1:

$$\tilde{g}(i\infty) = \alpha \neq 0; \tilde{g}(i) = 0; \tilde{g}(\tau) \text{ has a simple pole at } \tau = -1, \text{ with residue } 1.$$

We let $\tilde{f}_k(\tau) = \{\tilde{H}(\tau)\}^{k/2}$, where $\tilde{H}$ (replacing H) is defined by

$$\tilde{H}(\tau) = \frac{\tilde{g}'(\tau)}{\{\tilde{g}(\tau)\}^{\frac{1}{2}}\{g(\tau) - \alpha\}}. \tag{4.16}$$

As with the function $H(\tau)$ appearing in the proof of Theorem 4.1, $\tilde{H}(\tau)$ is analytic on $\mathcal{H}$ and zero-free on $\mathcal{H} \cup \{i\infty\}$; $\tilde{H}(\tau) \in M(\lambda, 2, 1)$ as well. It follows that $\tilde{f}_k(\tau) \in M(\lambda, k, 1)$ for arbitrary real k.

Define the sequence of functions

$$\tilde{g}_n(\tau) = \exp\{ni\,\tilde{g}(\tau)\}, \quad n \in \mathbb{Z}^+. \tag{4.17}$$

Then $\tilde{g}_n(\tau)$ is in $M(\lambda, 0, 1)$. Furthermore, for $\tau = -1 + iy$ and $y \to 0+$, $|\tilde{g}_n(\tau)| \sim e^{n/y}$. Clearly then the functions $g_n(\tau)$, $n \in \mathbb{Z}^+$, are linearly independent. On the other hand, by (4.16) and the asymptotic behavior of $\tilde{g}(\tau)$ near -1, $\tilde{H}(\tau) \sim y^{-1}$ (again, for $\tau = -1 + iy$ and $y \to 0+$), so that $\tilde{f}_k(\tau) \sim y^{-k/2}$. If we now put

$$\varphi_n(\tau) = \tilde{f}_k(\tau)\tilde{g}_n(\tau), \quad n \in \mathbb{Z}^+,$$

it follows directly that $\varphi_n(\tau) \in M(\lambda, k, 1)$, and, by Lemma 2.2, that the functions $\varphi_n(\tau)$ have property (4.15). This completes the proof of Theorem 4.2 for $\gamma = 1$.

The case $\gamma = -1$ can be treated exactly as in the proof of Theorem 4.1, with the role of $\tilde{f}_k(\tau)$ now assumed by $\tilde{h}_k(\tau) = \tilde{f}_k(\tau)\{\tilde{g}(\tau)\}^{\frac{1}{2}}$. This completes the proof of Theorem 4.2. □

Chapter 5

The case $\lambda < 2$

Recall from the proof of Lemma 3.1 that τ_λ denotes the lower left corner of $B(\lambda)$. Recall also that $\pi\theta = \pi - \arg(\tau_\lambda)$, so that $\cos(\pi\theta) = \lambda/2$.

Let $f \in M(\lambda, k, \gamma)$, $f \not\equiv 0$. Excluding the points τ_λ, $\tau_\lambda + \lambda$, i, and $i\infty$, let N denote the number of zeros of f on $\overline{B(\lambda)}$ counting multiplicities and with proper identification. That is to say, if $f(\tau_0) = 0$, where $\tau_0 \in \overline{B(\lambda)}$ and $\mathrm{Re}(\tau_0) = -\lambda/2$, then $f(\tau_0 + \lambda) = 0$, and N counts only one of the two zeros. Similarly, if $f(\tau_0) = 0$ with $\tau_0 \in \overline{B(\lambda)}$ and $|\tau_0| = 1$, then $f(-1/\tau_0) = 0$, and N counts only one of these two zeros. Let n_λ, n_i, and n_∞ denote the orders of the zeros of f at τ_λ, i, and $i\infty$, respectively. The order of the zero of f at $i\infty$ is measured in terms of $\exp(2\pi i\tau/\lambda)$.

Lemma 5.1. *We have*

$$N + n_\infty + \frac{1}{2}n_i + n_\lambda\theta = \frac{1}{2}k\left(\frac{1}{2} - \theta\right).$$

Proof. Let C denote the closed contour described below. The path C_1 is a horizontal segment from $\lambda/2 + iT$ to $-\lambda/2 + iT$, with $T > 0$ chosen large enough so that all of the zeros of f on $\overline{B(\lambda)}$, except for possibly $i\infty$, lie below C_1. This can be done because f is analytic at $i\infty$ in the local uniformizing variable $\exp(2\pi i\tau/\lambda)$. The path C_2 consists of that part of the line $x = -\lambda/2$, with indentations at the zeros of f (if any), from C_1 down to an arc C_{τ_λ} with center τ_λ and radius $\epsilon > 0$ passing from the line $x = -\lambda/2$ to $|\tau| = 1$. At this point on $|\tau| = 1$, C_3 begins and extends to C_i with semicircular indentations around zeros of f (if any). The arc C_i is a semicircular indentation about $\tau = i$. Lastly, $-C_4$ is the image of C_3 under the transformation $-1/\tau$, and C_5 is the image of C_2 under the transformation $\tau + \lambda$. If $C_{\tau_\lambda+\lambda}$ denotes the arc between C_4 and C_5, then, we define

$$C = C_1 + C_2 + C_{\tau_\lambda} + C_3 + C_i + C_4 + C_{\tau_\lambda+\lambda} + C_5.$$

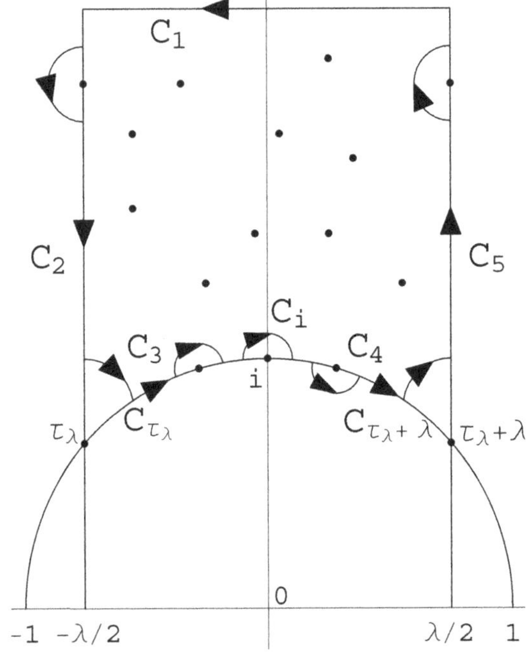

In this definition we choose C_i, C_{τ_λ}, and $C_{\tau_\lambda+\lambda}$ so that the portions of the discs bounded by these arcs and the boundary of $\overline{B(\lambda)}$ contain no zeros of f other than (possibly) the centers of the circular arcs about the points i, τ_λ, and $\tau_\lambda + \lambda$.

By the principle of the argument,

$$N = \frac{1}{2\pi i} \int_C \frac{f'(\tau)}{f(\tau)} d\tau. \tag{5.1}$$

Since $f'(\tau + \lambda)/f(\tau + \lambda) = f'(\tau)/f(\tau)$, it follows that

$$\frac{1}{2\pi i} \left\{ \int_{C_2} + \int_{C_5} \right\} \frac{f'(\tau)}{f(\tau)} d\tau = 0.$$

Since $f(-1/\tau) = \gamma(\tau/i)^k f(\tau)$, by logarithmic differentiation,

$$\frac{d}{d\tau} \log f(-1/\tau) = \frac{k}{\tau} + \frac{f'(\tau)}{f(\tau)}.$$

Hence,

$$\frac{1}{2\pi i}\left\{\int_{C_3}+\int_{C_4}\right\}\frac{f'(\tau)}{f(\tau)}d\tau \tag{5.2}$$
$$= \frac{1}{2\pi i}\int_{C_3}\frac{f'(\tau)}{f(\tau)}d\tau + \frac{1}{2\pi i}\int_{C_4}\frac{d}{d\tau}\log f(-1/\tau)d\tau - \frac{1}{2\pi i}\int_{C_4}\frac{k}{\tau}d\tau$$
$$= \frac{1}{2\pi i}\int_{C_3}\frac{f'(\tau)}{f(\tau)}d\tau + \frac{1}{2\pi i}\int_{-C_3}\frac{f'(\tau)}{f(\tau)}d\tau - \frac{1}{2\pi i}\int_{C_4}\frac{k}{\tau}d\tau$$
$$= -\frac{k}{2\pi i}\int_{C_4}\frac{d\tau}{\tau},$$

where, in the second integral on the right-hand side of (5.2), we have replaced τ by $-1/\tau$. Now,

$$-\frac{k}{2\pi i}\int_{C_4}\frac{d\tau}{\tau} = -\frac{k}{2\pi}\Delta_{C_4}(\arg), \tag{5.3}$$

where $\Delta_{C_4}(\arg)$ denotes the change in argument along C_4. Now as ϵ tends to 0, and the radii of the indentations tend to 0, the right-hand side of (5.3) tends to

$$\frac{k}{2\pi}\left(\frac{\pi}{2}-\pi\theta\right). \tag{5.4}$$

We next examine the contribution of the integral about C_i to the integral of (5.1). Write

$$f(\tau) = (\tau - i)^{n_i} f_1(\tau).$$

Thus,

$$f'(\tau) = n_i(\tau - i)^{n_i - 1} f_1(\tau) + (\tau - i)^{n_i} f_1'(\tau),$$

and

$$\frac{f'(\tau)}{f(\tau)} = \frac{n_i}{\tau - i} + \frac{f_1'(\tau)}{f_1(\tau)}.$$

Let δ denote the radius of the semicircular indentation C_i. Then,

$$\frac{1}{2\pi i}\int_{C_i}\frac{f'(\tau)}{f(\tau)}d\tau = \frac{n_i}{2\pi}\int_{\pi+\rho_1(\delta)}^{\rho_2(\delta)} d\varphi + \frac{1}{2\pi i}\int_{C_i}\frac{f_1'(\tau)}{f_1(\tau)}d\tau, \tag{5.5}$$

where $\rho_1(\delta)$ and $\rho_2(\delta)$ tend to 0 as δ tends to 0. Since $f_1(\tau)$ is analytic and nonzero in a neighborhood of $\tau = i$, if we let δ tend to 0, the right-hand side of (5.5) becomes

$$\frac{n_i}{2\pi}(-\pi) = -\frac{1}{2}n_i\,. \tag{5.6}$$

The contributions of C_{τ_λ} and $C_{\tau_\lambda+\lambda}$ to (5.1) are calculated in the same manner as above. Letting ϵ tend to 0, we find that the integrals

$$\frac{1}{2\pi i}\int_{C_{\tau_\lambda}} \frac{f'(\tau)}{f(\tau)}d\tau \quad \text{and} \quad \frac{1}{2\pi i}\int_{C_{\tau_\lambda}+\lambda} \frac{f'(\tau)}{f(\tau)}d\tau$$

each tend to

$$-\frac{1}{2}n_\lambda\theta, \tag{5.7}$$

since n_λ is also equal to the order of the zero of f at $\tau_\lambda + \lambda$, for $f(\tau) = f(\tau+\lambda)$.

Lastly, we examine the contribution of C_1 to (5.1) by letting C_1 tend to $i\infty$. Recall that we chose C_1 so that all finite zeros of f in $\overline{B(\lambda)}$ are below C_1. Put $z = \exp(2\pi i\tau/\lambda)$ and $f(\tau) = f_1(z)$. Thus, $f'(\tau) = f_1'(z)dz/d\tau$, and C_1 is transformed into a circle C_1^*, described negatively, of radius $\exp(-2\pi y/\lambda)$ about $z = 0$. Hence,

$$\frac{1}{2\pi i}\int_{C_1} \frac{f'(\tau)}{f(\tau)}d\tau = \frac{1}{2\pi i}\int_{C_1^*} \frac{f_1'(z)}{f_1(z)}dz = -n_\infty\,. \tag{5.8}$$

We now combine (5.4), (5.6), (5.7), and (5.8) with (5.1) to conclude that

$$N = \frac{1}{2}k\left(\frac{1}{2}-\theta\right) - \frac{1}{2}n_i - n_\lambda\theta - n_\infty\,,$$

and the proof is complete. □

Corollary 5.1. *Let*

$$m = 1 + \left[\frac{1}{2}k\left(\frac{1}{2}-\theta\right)\right].$$

Then,

$$\dim M(\lambda, k, \gamma) \le m.$$

Proof. By Lemma 5.1, $n_\infty < m$. Suppose that there exist $m+1$ linearly independent functions in $M(\lambda, k, \gamma)$. Then a suitable linear combination has a zero of order m at $i\infty$. (This suitable linear combination is achieved by solving a system of m linear homogeneous equations in m unknowns.) However, this is a contradiction to the fact that $n_\infty < m$. □

Thus, in contrast to Theorem 4.1, if $\lambda < 2$, $\dim M(\lambda, k, \gamma)$ is finite and Corollary 5.1 provides an upper bound for $\dim M(\lambda, k, \gamma)$.

Example 5.1. Let $\lambda = \sqrt{3}$, so that $\cos(\pi/6) = \sqrt{3}/2$ and $\theta = 1/6$. Let $k = \gamma = 1$. Then,

$$m = 1 + \left[\frac{1}{2}\left(\frac{1}{2} - \frac{1}{6}\right)\right] = 1.$$

Hence, there is at most one modular form in $M(\sqrt{3}, 1, 1,)$ or $M_0(\sqrt{3}, 1, 1)$, i.e., there exists at most one Dirichlet series of signature $(\sqrt{3}, 1, 1,)$. However, there is known to exist such a series, namely the Dedekind zeta function $\zeta_K(s)$ for $Q(\sqrt{-3})$,

$$\zeta_K(s) = \frac{1}{6} \sum_{m,n=-\infty}^{\infty}{}' \{Q(m,n)\}^{-s} = \frac{1}{6} Z(s, Q),$$

where $Q(m, n) = m^2 + mn + n^2$.

Before proceeding further we must make some remarks about elliptic linear fractional transformations.

Definition 5.1. Let a, b, c, and d be real with $ad - bc = 1$. Then we say that $V(\tau) = (a\tau + b)/(c\tau + d)$ is *elliptic if* V has two nonreal fixed points τ_0 and $\bar{\tau}_0$, where $\mathrm{Im}(\tau_0) > 0$.

We note that V is elliptic if and only if

$$(d - a)^2 + 4bc < 0,$$

or

$$(d - a)^2 + 4(ad - 1) < 0,$$

or

$$(a + d)^2 < 4.$$

Thus, V is elliptic if and only if the roots of

$$x^2 - (a + d)x + 1 = 0 \tag{5.9}$$

are nonreal. Since (5.9) is equivalent to

$$\begin{vmatrix} a-x & b \\ c & d-x \end{vmatrix} = 0,$$

the roots of (5.9) are the *eigenvalues* of V. Note that, by (5.9), the eigenvalues of an elliptic linear fractional transformation (or its corresponding matrix) with real entries have absolute value 1.

Lemma 5.2. *Let V be elliptic with a fixed point $\tau_0 \in \mathcal{H}$. Suppose that there exists a nonconstant function f, analytic on $\mathcal{H}$, such that*

$$f(\tau) = f(\tau + \lambda)$$

and

$$f(V\tau) = \epsilon \left(\frac{c\tau + d}{i}\right)^k f(\tau),$$

for some constants λ, $k > 0$, and ϵ, $|\epsilon| = 1$. Then, if ρ is an eigenvalue of V, ρ is a root of unity. Consequently, V has finite order.

Proof. First, observe that ρ is a solution of the equations

$$x\tau_0 = a\tau_0 + b \quad \text{and} \quad x = c\tau_0 + d,$$

for if ρ is such a solution, then

$$\begin{vmatrix} a-\rho & b \\ c & d-\rho \end{vmatrix} = 0.$$

Set

$$\hat{t}(\tau) = \frac{\tau - \tau_0}{\tau - \bar{\tau}_0}.$$

(Compare this with $t_1(\tau)$, used to define the local uniformizing variable at i. See Chapter 4, just before (4.4).) By a direct calculation,

$$V(\tau) - V(\tau_0) = \frac{\tau - \tau_0}{(c\tau + d)(c\tau_0 + d)}.$$

Thus, as τ_0 and $\bar{\tau}_0$ are fixed points of V,

$$\hat{t}(V(\tau)) = \frac{V(\tau) - \tau_0}{V(\tau) - \bar{\tau}_0} = \frac{V(\tau) - V(\tau_0)}{V(\tau) - V(\bar{\tau}_0)} = \frac{c\bar{\tau}_0 + d}{c\tau_0 + d}\frac{\tau - \tau_0}{\tau - \bar{\tau}_0} = \bar{\rho}^2 \hat{t}(\tau),$$

since $c\bar{\tau}_0 + d = \bar{\rho}$ and $|\rho| = 1$. Hence, replacing $\bar{\rho}$ by ρ, we have shown that ρ is a root of

$$\hat{t}(V(\tau)) = \rho^2 \hat{t}(\tau). \tag{5.10}$$

Let

$$g(\tau) = (\tau - \bar{\tau}_0)^k, \quad \tau \in \mathcal{H}.$$

Then, since $\bar{\tau}_0$ is fixed by V,

$$\begin{aligned} g(V\tau) &= \{V(\tau) - V(\bar{\tau}_0)\}^k \\ &= \left\{ \frac{a\tau + b}{c\tau + d} - \frac{a\bar{\tau}_0 + b}{c\bar{\tau}_0 + d} \right\}^k \\ &= \left\{ \frac{\tau - \bar{\tau}_0}{(c\tau + d)(c\bar{\tau}_0 + d)} \right\}^k \\ &= \frac{\eta g(\tau)}{\{-i(c\tau + d)\}^k}, \end{aligned}$$

where η is a constant. If we let $\tau = \tau_0$, then

$$g(\tau_0) = g(V\tau_0) = \frac{\eta g(\tau_0)}{\{-i(c\tau_0 + d)\}^k}, \tag{5.11}$$

and so

$$\eta = \{-i(c\tau_0 + d)\}^k, \tag{5.12}$$

since $g(\tau_0) \neq 0$.

Next, let

$$h(\hat{t}) = f(\tau)g(\tau) = f\left(\frac{\bar{\tau}_0\hat{t} - \tau_0}{\hat{t} - 1}\right) g\left(\frac{\bar{\tau}_0\hat{t} - \tau_0}{\hat{t} - 1}\right).$$

Then, from the hypotheses, (5.11), and (5.12),

$$\begin{aligned} h(\hat{t}(V\tau)) &= f(V\tau)g(V\tau) \\ &= \epsilon \left(\frac{c\tau + d}{i}\right)^k f(\tau) \left(\frac{i}{c\tau + d}\right)^k \eta g(\tau) \\ &= \epsilon\eta h(\hat{t}). \end{aligned} \tag{5.13}$$

Since both f and g are analytic on $\mathcal{H}$, $h(\hat{t})$ has a Maclaurin series expansion

$$h(\hat{t}) = \sum_{n=0}^{\infty} a_n \hat{t}^n.$$

Now, from (5.10) and (5.13),

$$h(\rho^2 \hat{t}(\tau)) = h(\hat{t}(V\tau)) = \epsilon\eta h(\hat{t}),$$

or

$$\sum_{n=0}^{\infty} a_n \rho^{2n} \hat{t}^n = \epsilon\eta \sum_{n=0}^{\infty} a_n \hat{t}^n. \tag{5.14}$$

By hypothesis, f is periodic, so it is clear that $h(\hat{t})$ cannot reduce to a polynomial in $\hat{t}$. Assume then that $a_{n_1}, a_{n_2} \neq 0$, with $n_1 \neq n_2$. Thus,

$$\rho^{2n_1} = \epsilon\eta \quad \text{and} \quad \rho^{2n_2} = \epsilon\eta,$$

or

$$\rho^{2(n_1 - n_2)} = 1,$$

and the result is proved. □

Remark 5.1. By Corollary 14 of [59, p. 520], the existence of a nonconstant function f satisfying the hypotheses of Lemma 5.2 implies that the group $\langle S_\lambda, V \rangle$ is discrete. From this, in turn, it follows that V has finite order [73, p. 91], [74, p. 15]. On the other hand, our proof of Lemma 5.2 derives the same result more simply and directly, however without placing it within its natural setting of discrete groups.

Corollary 5.2. *Let $f \in M(\lambda, k, \gamma)$ and let n_i be the order of the zero of f at $\tau = i$. Then*

$$\gamma = (-1)^{n_i}.$$

Proof. Let $V\tau = T\tau = -1/\tau$, an elliptic transformation with fixed points $\pm i$. In the notation of Chapter 4, let

$$t_1(\tau) = \frac{\tau - i}{\tau + i};$$

by the proof of Lemma 5.2,

$$t_1(T\tau) = \frac{-1/\tau - i}{-1/\tau + i} = -\frac{\tau - i}{\tau + i}.$$

Thus, by comparison with (5.10), ρ is an eigenvalue, where $\rho^2 = -1$. Since

$$f(-1/\tau) = \gamma(\tau/i)^k f(\tau),$$

$\gamma = \epsilon$ in the notation of the previous lemma. From (5.12), $\eta = 1$. Now, in (5.14) $a_{n_i} \neq 0$, for f has a zero at $\tau = i$ of order precisely n_i. Thus, equating coefficients of t^{n_i} in (5.14), we find that

$$(-1)^{n_i} = \rho^{2n_i} = \epsilon\eta = \gamma,$$

and the proof is complete. □

Theorem 5.1. *If* $\dim M(\lambda, k, \gamma) \neq 0$, *then* θ *and* k *are rational.*

Proof. Let

$$V(\tau) = T_1 T_3(\tau) = -1/(\tau + \lambda).$$

Then, writing $\tau_\lambda = -\frac{1}{2}\lambda + iy_0$, with $y_0 = \sqrt{1 - \lambda^2/4}$, we see that

$$V(\tau_\lambda) = -\frac{1}{\frac{1}{2}\lambda + iy_0} = \frac{-\frac{1}{2}\lambda + iy_0}{|\frac{1}{2}\lambda + iy_0|^2} = -\frac{1}{2}\lambda + iy_0 = \tau_\lambda\,,$$

i.e., τ_λ is a fixed point of V, and V is elliptic. In the notation of Definition 5.1, $a + d = \lambda$, and so, by (5.9), the eigenvalues of V are roots of

$$x^2 - \lambda x + 1 = 0.$$

Thus, an eigenvalue is

$$\rho = \frac{\lambda + \sqrt{\lambda^2 - 4}}{2} = \frac{1}{2}\lambda + \sqrt{\lambda^2/4 - 1}$$
$$= \cos \pi\theta + i \sin \pi\theta = e^{\pi i \theta}.$$

By Lemma 5.2, ρ is a root of unity, and hence θ is rational. Also, by Lemma 5.1,

$$N + n_\infty + \frac{1}{2}n_i + n_\lambda \theta = \frac{1}{2}k\left(\frac{1}{2} - \theta\right),$$

and it follows that k is rational. □

Remark 5.2. It has been noted in other contexts (see, e.g., [64, Theorem 7]) that $k \in \mathbb{Q}$ follows from the periodicity of $f \in M(\lambda, k, \gamma)$. Were we to assume that $f(\tau + \lambda) = \epsilon f(\tau)$, with ϵ not a root of unity, then k would in fact be irrational. Furthermore, if $\lambda \geq 2$, then the rationality of k does not follow from $f(\tau+\lambda)=f(\tau)$, since, in this case, there is no nontrivial relation in $G(\lambda)$ involving both S_λ and T. (For example, let $f(\tau)=\theta^s(\tau)$, with irrational real s. Then the weight $k=s/2\notin\mathbb{Q}$, but $f(\tau+2)=f(\tau)$. Of course, since $\lambda=2$ here, Theorem 5.1 does not apply to this case. See Chapters 1 and 4.)

Example 5.2. We briefly indicate what our results so far yield for the modular group $G(1)$. Here, $\cos\pi\theta=\frac{1}{2}$, and so $\theta=\frac{1}{3}$. If $f\in M(1,k,\gamma)$, by Lemma 5.1, we find that

$$n_i/2+n_1/3\equiv k/12 \pmod 1,$$

which implies that

$$3n_i/2+n_1\equiv k/4 \pmod 1,$$

or

$$n_i/2\equiv k/4 \pmod 1.$$

Hence, k is an even positive integer. By Corollary 5.2, this last congruence also shows that

$$\gamma=(-1)^{n_i}=(-1)^{k/2}.$$

Furthermore, from Lemma 5.1,

$$N+n_\infty+n_i/2+n_1/3=k/12,$$

so that

$$k/12\geq n_1/3.$$

Hence, $k\geq 4$ as well, since we have assumed that $k>0$, and, furthermore, $n_1\geq 1$ when $k=2$. The latter observation follows, since by definition of $M(1,k,\gamma)$,

$$\gamma(\tau_1/i)^2 f(\tau_1)=f(-1/\tau_1)=f(\tau_1+1)=f(\tau_1)$$

(recall that $\tau_1=\frac{1}{2}(-1+i\sqrt{3})$), while $\gamma(\tau_1/i)^2\neq 1$. Lastly, from Corollary 5.1,

$$\dim M(1,k,(-1)^{k/2})\leq 1+[k/12].$$

We have already shown that θ must be rational if $\dim M(\lambda, k, \gamma) \neq 0$. The next theorem shows more yet.

Theorem 5.2. *If* $\dim M(\gamma, k, \gamma) \neq 0$, *then* $\theta = 1/q$, *where* $q \geq 3$ *and* $q \in \mathbb{Z}$.

Proof. Let $\theta = p/q$, where $(p, q) = 1$ and $p \geq 2$. The idea of the proof is to construct a transformation V such that V is elliptic but also such that its eigenvalues are not roots of unity. However, by Lemma 5.2 this is a contradiction, and so the proof would be complete.

By (3.1), if $\theta = p/q$,

$$(T_1T_3)^n = \begin{bmatrix} 0 & -1 \\ 1 & \lambda \end{bmatrix}^n = \frac{1}{\sin(\pi p/q)} \begin{bmatrix} \sin\{\pi p(1-n)/q\} & -\sin(\pi pn/q) \\ \sin(\pi pn/q) & \sin\{\pi p(1+n)/q\} \end{bmatrix}.$$

Let

$$V = S_\lambda (T_1T_3)^n,$$

for a certain choice of $n \in \mathbb{Z}$ to be given shortly. Note that

$$\begin{aligned} \operatorname{tr}(V) &= \frac{\sin\{\pi p(1-n)/q\} + \lambda \sin(\pi pn/q) + \sin\{\pi p(n+1)/q\}}{\sin(\pi p/q)} \\ &= \frac{2\sin(\pi p/q)\cos(\pi pn/q) + 2\cos(\pi p/q)\sin(\pi pn/q)}{\sin(\pi p/q)} \\ &= \frac{2\sin\{\pi p(n+1)/q\}}{\sin(\pi p/q)}. \end{aligned}$$

Since $(p, q) = 1$, we can choose n so that

$$(n+1)p \equiv 1 \pmod{q}.$$

Hence,

$$|\operatorname{tr}(V)| = 2\left|\frac{\sin(\pi/q)}{\sin(\pi p/q)}\right| < 2,$$

as $p \geq 2$. (Note that since $\cos \pi\theta > 0$, $\theta < \frac{1}{2}$, and so the case $p = q - 1$ is impossible.) Now the eigenvalues of V are the roots of

$$\rho^2 - (\operatorname{tr}(V))\rho + 1 = 0,$$

and since $|\operatorname{tr}(V)| < 2$, V is elliptic.

Let $r = \operatorname{tr}(V)$. We now replace r by an algebraic conjugate r' given by

$$r' = \frac{2\sin\{\pi p\ell(n+1)/q\}}{\sin(\pi p\ell/q)},$$

for a suitable integer ℓ. Then the solutions ρ_1' and ρ_2' of

$$\rho'^2 - r'\rho' + 1 = 0$$

are algebraic conjugates of ρ. We choose ℓ so that $p\ell \equiv 1 \pmod{q}$, which is possible since $(p, q) = 1$. Hence,

$$|r'| = 2\left|\frac{\sin(\pi\ell/q)}{\sin(\pi/q)}\right| \geq 2.$$

In fact, $|r'| > 2$, for if $\ell \equiv \pm 1 \pmod{q}$, then

$$|r'| = 2\left|\frac{\sin(\pi/q)}{\sin(\pi p/q)}\right| = |r| < 2,$$

which is a contradiction. Thus, ρ' is real, and at least one of the two values ρ_1', ρ_2' has modulus greater than 1 since $|\rho_1' + \rho_2'| = |r'| > 2$. We are forced to conclude that a root of unity ρ has a conjugate ρ' with $|\rho'| > 1$, which is impossible. □

Note that if $p = 1$ the proof fails, for then $|r| = 2$.

To summarize our work so far, if there exists a nontrivial function $f \in M(\lambda, k, \gamma)$, then

$$\gamma = (-1)^{n_i}, \tag{5.15}$$

$$N + n_\infty + \frac{1}{2}n_i + \frac{n_\lambda}{q} = \frac{k(q-2)}{4q}, \qquad q \geq 3,\ q \in \mathbb{Z}, \tag{5.16}$$

and

$$\lambda = 2\cos(\pi/q), \qquad q \geq 3, \qquad q \in \mathbb{Z}, \quad \text{i.e.,} \quad \lambda \in T_A.$$

The derivation of Theorem 5.2 proves more than is claimed in the statement. In fact, the essential feature of the proof is an application of Lemma 5.2, and in the proof of the latter we require only that f is a nonconstant, periodic, meromorphic function in $\mathcal{H}$ satisfying the stated transformation law for some elliptic $V = \left(\begin{smallmatrix} * & * \\ c & d \end{smallmatrix}\right) \in G(\lambda)$. Thus we can reformulate Theorem 5.2 in a stronger fashion.

Theorem 5.3. *Suppose there exists a nonconstant function f, meromorphic in $\mathcal{H}$ and satisfying the transformation laws $f(\tau+\lambda) = f(\tau)$ and*

$$f(V\tau) = \epsilon \left(\frac{c\tau + d}{i}\right)^k f(\tau),$$

for some elliptic $V = \left(\begin{smallmatrix} * & * \\ c & d \end{smallmatrix}\right) \in G(\lambda)$, with constants $\lambda > 0$, k real, and $\epsilon \neq 0$. Then if $\lambda < 2$, it follows that $\lambda = 2\cos(\pi/q)$, where $q \in \mathbb{Z}$ and $q \geq 3$.

Note that in Lemma 5.2 the assumption on the multiplier ϵ is $|\epsilon| = 1$, whereas in Theorem 5.3 we assume only that $\epsilon \neq 0$. But, if $V \in G(\lambda)$ is elliptic and $\epsilon = \epsilon(V) \neq 0$, it follows from Lemma 5.2 that V has finite order, and this, together with a calculation, implies $|\epsilon(V)| = 1$. (Note that the proof of Lemma 5.2 uses only $\epsilon \neq 0$, rather than $|\epsilon| = 1$.)

The significance of this slightly stronger version emerges when we combine it with the final theorem in [59]: *If Γ is a linear fractional group preserving $\mathcal{H}$ and Γ contains $S_\lambda = \left(\begin{smallmatrix} 1 & \lambda \\ 0 & 1 \end{smallmatrix}\right)$ for some $\lambda > 0$, then Γ is discrete if and only if Γ supports a nonconstant automorphic form (of some real weight).*

By Theorem 5.3 and Theorem 5.5 (below), when $\lambda < 2$, $G(\lambda)$ supports a nonconstant automorphic form if and only if $\lambda = 2\cos(\pi/q)$, with $q \in \mathbb{Z}$ and $q \geq 3$. In summary, then, we can state the following theorem.

Theorem 5.4. *Let $\lambda < 2$. Then $G(\lambda)$ is discrete if and only if $\lambda = 2\cos(\pi/q)$, with $q \in Z$ and $q \geq 3$.*

Theorem 5.5. *Let $\lambda \in T_A$. Then there exist functions f_λ, f_i, and $f_\infty \in M(\lambda, k, \gamma)$ such that each has a simple zero at τ_λ, i, and $i\infty$, respectively, and no other zeros. Here, γ is given by (5.15), and k is determined in each case from (5.16). Thus, $f_\lambda \in M(\lambda, 4/(q-2), 1)$, $f_i \in M(\lambda, 2q/(q-2), -1)$, and $f_\infty \in M(\lambda, 4q/(q-2), 1)$.*

Note that there exists at most one function in each case satisfying the conditions of the theorem. For if there were two such functions, then a suitable linear combination would have a zero at $i\infty$ (or a double zero in the third case of f_∞). This however would be incompatible with (5.16). When we say that, for example, f_λ has only one zero, we mean, of course, up to equivalence.

Proof of Theorem 5.5. We proceed as in the proof of Theorem 4.1. By the Riemann mapping theorem there exists a function $g(\tau)$ that maps the simply connected region $B(\lambda)$ one-to-one and conformally onto $\mathcal{H}$. As in the proof of Theorem 4.1, the mapping is determined uniquely if we specify the values of g at three points on the boundary of $B(\lambda)$. We shall require that

$$g(\tau_\lambda) = 0,$$
$$g(i) = 1,$$

and

$$g(i\infty) = \infty.$$

We note that g maps the arc of the unit circle from τ_λ to i onto $[0, 1]$. The arc of the unit circle from i to $\tau_\lambda + \lambda$ is mapped onto $[1, a_0]$ for some real number $a_0 > 1$. The right-hand edge of $B(\lambda)$ is mapped onto $[a_0, \infty)$, and the left-hand edge of $B(\lambda)$ is mapped onto $(-\infty, 0]$. As before, we continue g analytically to $\mathcal{H}$ by defining

$$g(T_i\tau) = \overline{g(\tau)}, \quad i = 1, 2, 3,$$

employing repeated reflections, and using the Schwarz reflection principle. Because $B(\lambda)$ is a fundamental region for $G(\lambda)$, $g(\tau)$ is well defined. Furthermore, $g(\tau)$ is invariant under transformations of $G(\lambda)$ and analytic on $\mathcal{H}$.

We now examine g in neighborhoods of $\tau = i$, τ_λ, and $i\infty$. We first consider g in a neighborhood of i. Let

$$t_1(\tau) = (\tau - i)/(\tau + i).$$

From (4.4), $t_1(-1/\tau) = -t_1(\tau)$. As before, $t(\tau) = \{t_1(\tau)\}^2$ is a local uniformizing variable for g in a neighborhood of i. If we put $g_1(t) = g(\tau)$, then $g_1(t)$ is one-to-one in a neighborhood of $t = 0$, and

$$g_1(t) = 1 + \sum a_n t^n,$$

with $a_1 \neq 0$. In other words, $g(\tau)$ has a double value 1 at $\tau = i$, in the standard planar local variable at i, that is, $\tau - i$. (As we have shown, of course, in the local variable $t = \{(\tau - i)/(\tau + i)\}^2$ with respect to $G(\lambda)$, $g(\tau)$ has a *simple* value 1.)

Near $\tau = \tau_\lambda$ we put

$$t_2(\tau) = (\tau - \tau_\lambda)/(\tau - \bar{\tau}_\lambda).$$

Now,

$$t_2(T_1T_3\tau) = \frac{-1/(\tau+\lambda) - \tau_\lambda}{-1/(\tau+\lambda) - \bar{\tau}_\lambda} = \frac{\tau_\lambda(\tau + \tau_\lambda^{-1} + \lambda)}{\bar{\tau}_\lambda(\tau + \tau_\lambda^{-1} + \lambda)}.$$

Note that $\tau_\lambda = \exp\{i\pi(1 - 1/q)\}$, and so $\tau_\lambda/\bar{\tau}_\lambda = \exp(-2\pi i/q)$. Also,

$$\frac{1}{\tau_\lambda} + \lambda = \frac{1}{-\frac{1}{2}\lambda + iy} + \lambda = -\frac{1}{2}\lambda - iy + \lambda = -\tau_\lambda.$$

Thus,

$$t_2(T_1T_3\tau) = t_2(-1/(\tau+\lambda)) = e^{-2\pi i/q}\,\frac{\tau - \tau_\lambda}{\tau - \bar{\tau}_\lambda} = e^{-2\pi i/q}t_2(\tau).$$

Now, in analogy with our earlier argument (in a neighborhood of $\tau = i$), we observe that $g(\tau)$ can be rewritten as a function of t_2, once again since $\tau \to t_2(\tau)$ is one-to-one on the Riemann sphere; write $g(\tau) = \tilde{g}(t_2)$. Furthermore, if we put $\tau' = -1/(\tau + \lambda)$, then $t_2(\tau') = e^{-2\pi i/q}t_2(\tau)$ and $g(\tau') = g(\tau)$ together imply that $\tilde{g}_2(t_2)$ is actually a function of $t(\tau) = \{t_2(\tau)\}^q$, that is,

$$g(\tau) = \tilde{g}(t_2) = g_2(t_2^q) = g_2(t), \qquad t = t_2^q.$$

Since $g_2(0) = g(\tau_\lambda) = 0$ and $g_2(t)$ is one-to-one in a neighborhood of $t = 0$, we see that

$$g_2(t) = \sum b_n t^n,$$

with $b_1 \neq 0$. Thus, $g(\tau)$ has a simple zero at τ_λ in the local variable $t = t_2^q = \{(\tau - \tau_\lambda)/(\tau + \bar{\tau}_\lambda)\}^q$ with respect to $G(\lambda)$. (This means that $g(\tau)$ has a zero of order q in the standard local variable $\tau - \tau_\lambda$.)

Next, put $t = \exp(2\pi i\tau/\lambda)$. The argument used in Chapter 4 to derive (4.5) shows that we can write $g(\tau) = g_3(t)$, a single-valued function of t. Since g is one-to-one on $B(\lambda)$, it follows that $g_3(t)$ is one-to-one in a neighborhood of $t = 0$. Since also $g(i\infty) = \infty$, $g_3(t)$ has a simple pole at $t = 0$, or $g(\tau)$ has a simple pole at $i\infty$, where the pole is measured in terms of $t = \exp(2\pi i\tau/\lambda)$. Furthermore, if

$$g(\tau) = a_{-1}e^{-2\pi i\tau/\lambda} + a_0 + a_1e^{2\pi i\tau/\lambda} + \cdots,$$

then

$$g'(\tau) = a_{-1}(-2\pi i/\lambda)e^{-2\pi i\tau/\lambda} + a_1(2\pi i/\lambda)e^{2\pi i\tau/\lambda} + \cdots,$$

and since $a_{-1} \neq 0$, $g'(\tau)$ also has a simple pole at $i\infty$.

We are now ready to construct the functions f_λ, f_i, and f_∞. Define

$$f_\lambda(\tau) = \left\{ \frac{\{g'(\tau)\}^2}{g(\tau)(g(\tau)-1)} \right\}^{1/(q-2)},$$

$$f_i(\tau) = \left\{ \frac{\{g'(\tau)\}^q}{\{g(\tau)\}^{q-1}(g(\tau)-1)} \right\}^{1/(q-2)},$$

and

$$f_\infty(\tau) = \left\{ \frac{\{g'(\tau)\}^{2q}}{\{g(\tau)\}^{2q-2}(g(\tau)-1)^q} \right\}^{1/(q-2)}.$$

We must show that these functions have the properties enunciated in the theorem.

Since g has a zero of order q at τ_λ, $(g')^2$ has a zero of order $2(q-1)$ at τ_λ. Thus, $(g')^2/g$ has a zero of order $q-2$ at τ_λ. Also, $g(\tau_\lambda) - 1 \neq 0$, since g is one-to-one on $B(\lambda)$ and $g(i) = 1$. Hence, f_λ has a simple zero at $\tau = \tau_\lambda$. Now,

$$\frac{\{g'(\tau)\}^2}{g(\tau)(g(\tau)-1)} = (\tau - \tau_\lambda)^{q-2} g_1(\tau), \tag{5.17}$$

where g_1 is analytic at τ_λ and $g_1(\tau_\lambda) \neq 0$. Thus, we can define an analytic $1/(q-2)$th root of (5.17) in a neighborhood of τ_λ. Since $g(\tau) - 1$ has a double zero at $\tau = i$, and $g'(\tau)$ has a simple zero at i, f_λ is analytic and nonzero at $\tau = i$. Lastly, note that $(g')^2$ has a double pole at $i\infty$, and $g(\tau)(g(\tau)-1)$ has a double pole at $i\infty$. Hence, f_λ is defined and not zero at $i\infty$. In conclusion, f_λ has only one zero on $\mathcal{H} \cup \{i\infty\}$, a simple zero at τ_λ. We can define an analytic $1/(q-2)$th root in a neighborhood of every point of $\mathcal{H}$. Hence, since $\mathcal{H}$ is simply connected, f_λ is analytic on $\mathcal{H}$ by the monodromy theorem.

We consider next f_i. Observe that $(g')^q$ has a zero of order q at i, and $g(\tau) - 1$ has a zero of order 2 at i. Hence, f_i has a simple zero at $\tau = i$. Also, $(g')^q$ has a zero at τ_λ of order $q(q-1)$, and g^{q-1} has a zero at τ_λ of order $(q-1)q$. Since $g(\tau_\lambda) - 1 \neq 0$, we conclude that f_i is analytic and nonzero at $\tau = \tau_\lambda$. Lastly, $(g')^q$ has a pole of order q at $i\infty$. Also, $g^{q-1}(g-1)$ has a pole of order q at $i\infty$. Hence, f_i is defined and not zero at $i\infty$. We conclude that f_i has but one zero on $\mathcal{H} \cup \{i\infty\}$, a simple zero at $\tau = i$. Just as with f_λ, we may define an analytic $1/(q-2)$th root in a neighborhood of every point of $\mathcal{H}$ and conclude as before that f_i is analytic on $\mathcal{H}$.

Lastly, we consider f_∞. Note that $(g')^{2q}$ has a pole at $i\infty$ of order $2q$. Also, $g^{2q-2}(g-1)^q$ has a pole at $i\infty$ of order $3q-2$. Thus, f_∞ has a simple zero at $i\infty$. Furthermore, $(g')^{2q}$ has a zero of order $2q(q-1)$ at τ_λ. Also, g^{2q-2} has a zero at τ_λ of order $(2q-2)q$. Hence, f_∞ is analytic and nonzero at τ_λ. Now, $(g')^{2q}$ has a zero of order $2q$ at i and $(g-1)^q$ has a zero of order $2q$ at i. Hence, f_∞ is analytic and nonzero at i. Thus, f_∞ has only one zero on $\mathcal{H} \cup \{i\infty\}$, a simple one at $i\infty$. As above, f_∞ is analytic on $\mathcal{H}$.

To conclude the proof of the theorem we must check that (ii) of Definition 2.2 is satisfied with the proper value of k. Since $g(-1/\tau) = g(\tau)$, we have

$$\frac{d}{dt}g(-1/\tau) = \tau^2 g'(\tau).$$

First,

$$\begin{aligned} f_\lambda(-1/\tau) &= \left\{ \frac{\{g'(-1/\tau)\}^2}{g(-1/\tau)(g(-1/\tau)-1)} \right\}^{1/(q-2)} \\ &= \left\{ \frac{\tau^4\{g'(\tau)\}^2}{g(\tau)(g(\tau)-1)} \right\}^{1/(q-2)} \\ &= \epsilon(\tau/i)^{4/(q-2)} f_\lambda(\tau), \end{aligned}$$

where $|\epsilon| = 1$. Since $f_\lambda(i) \neq 0$, by letting $\tau = i$ above, we find that $\epsilon = 1$. Hence, $f_\lambda \in M(\lambda, 4/(q-2), 1)$.

Secondly,

$$\begin{aligned} f_i(-1/\tau) &= \left\{ \frac{\{g'(-1/\tau)\}^q}{\{g(-1/\tau)\}^{q-1}(g(-1/\tau)-1)} \right\}^{1/(q-2)} \\ &= \left\{ \frac{\tau^{2q}\{g'(\tau)\}^q}{\{g(\tau)\}^{q-1}(g(\tau)-1)} \right\}^{1/(q-2)} \\ &= \epsilon(\tau/i)^{2q/(q-2)} f_i(\tau), \end{aligned}$$

where $|\epsilon| = 1$. Since f_i has a simple zero at $\tau = i$, write

$$f_i(\tau) = (\tau - i) f_1(\tau).$$

Then,

$$f_i(-1/\tau) = (-1/\tau - i) f_1(-1/\tau) = -\frac{i}{\tau}(\tau - i) f_1(-1/\tau).$$

Hence,

$$\frac{f_i(-1/\tau)}{\tau - i} = \epsilon \left(\frac{\tau}{i}\right)^{2q/(q-2)} \frac{f_i(\tau)}{\tau - i},$$

which implies that

$$-\frac{i}{\tau} f_1(-1/\tau) = \epsilon \left(\frac{\tau}{i}\right)^{2q/(q-2)} f_1(\tau).$$

Since $f_1(i) \neq 0$, by putting $\tau = i$ above, we find that $\epsilon = -1$. Hence, $f_i \in M(\lambda, 2q/(q-2), -1)$.

Thirdly,

$$\begin{aligned} f_\infty(-1/\tau) &= \left\{ \frac{\{g'(-1/\tau)\}^{2q}}{\{g(-1/\tau)\}^{2q-2}(g(-1/\tau)-1)^q} \right\}^{1/(q-2)} \\ &= \left\{ \frac{\tau^{4q}\{g'(\tau)\}^{2q}}{\{g(\tau)\}^{2q-2}(g(\tau)-1)^q} \right\}^{1/(q-2)} \\ &= \epsilon(\tau/i)^{4q/(q-2)} f_\infty(\tau), \end{aligned}$$

where $|\epsilon| = 1$. Since $f_\infty(i) \neq 0$, we can let $\tau = i$ above to conclude that $\epsilon = 1$. Hence, $f_\infty \in M(\lambda, 4q/(q-2), 1)$. □

We are now able, in the next theorem, to give more precise information than that contained in Corollary 5.1, Theorem 5.1, and Theorem 5.2.

Theorem 5.6. *If $\lambda \notin T_A$, then $\dim M(\lambda, k, \gamma) = 0$. If $\lambda = 2\cos(\pi/q) \in T_A$, then for nontrivial $f \in M(\lambda, k, \gamma)$, the weight k has the form*

$$k = \frac{4m}{q-2} + 1 - \gamma, \tag{5.18}$$

where $m \geq 1$ is an integer. Furthermore,

$$\dim M(\lambda, k, \gamma) = 1 + \left[\frac{m + (\gamma - 1)/2}{q}\right]. \tag{5.19}$$

Remark 5.3. 1. Since $m \geq 1$ in formula (5.18), the case $m = 0$ is not included here in (5.19). Nevertheless, the formula (5.19) does hold when $m = 0$ in (5.18).

To see this, note first that (5.18) with $m = 0$ implies that $k = 1 - \gamma = 0$ or 2, according as $\gamma = 1$ or -1. We show that in either event the existence of nonconstant $f \in M(\lambda, 1-\gamma, \gamma)$ is ruled out by the relation (5.16).

Suppose that $\gamma = 1$ and $k = 0$. Assume $f \in M(\lambda, 0, 1)$ and suppose $f(z_0) = c$ for some $z_0 \in \mathcal{H}$. Then letting $g(z) = f(z) - c$, we conclude that

$g \in M(\lambda, 0, 1)$ and $g(z_0) = 0$. But an application of (5.16) to g implies that g is zero-free in $\mathcal{H} \cup \{i\infty\}$. This is a contradiction unless $g(z) \equiv 0$, and so $f(z) \equiv c$. We conclude that $\dim M(\lambda, 0, 1) = 1$.

Suppose next that $\gamma = -1$ and $k = 2$. Then assumption (ii) in the definition of $M(\lambda, 2, -1)$ implies that $f(i) = -f(i)$, so $f(i) = 0$ and $n_i \geq 1$. But, (5.16) in this case yields

$$\frac{1}{2}n_i \leq N + n_\infty + n_{i/2} + n_{\lambda/q} = \frac{q-2}{2q} < \frac{1}{2}.$$

This contradiction implies that $f(z) \equiv 0$, so that $\dim M(\lambda, 2, -1) = 0$.

On the other hand, when $m = 0$ and $\gamma = 1$, the right-hand side of (5.19) equals 1, while the right-hand side of (5.19) is $1 + [-1/q] = 0$ when $m = 0$ and $\gamma = -1$. Thus (5.19) is valid when $m = 0$ in (5.18).

2. An application of (5.16) as in Remark 5.3. 1. (above) leads to the observation that $\dim M(\lambda, k, \gamma) = 0$ if $k = 0$, $\gamma = -1$, and if $k < 0$.

Proof of Theorem 5.6. First, suppose that $f \in M(\lambda, k, 1)$. From (5.15), n_i is even. From (5.16),

$$\frac{k(q-2)}{4} = m,$$

where m is a positive integer. Therefore,

$$k = \frac{4m}{q-2} = mk_\lambda,$$

where $k_\lambda = 4/(q-2)$, the minimum value of the parameter (i.e., the weight) k of nontrivial $f \in M(\lambda, k, 1)$ and the value k for the invariant function f_λ of Theorem 5.5. This proves (5.18) for $\gamma = 1$.

Proceeding with the case $\gamma = 1$, we put

$$f_1(\tau) = f(\tau) - \alpha f_\lambda^m(\tau),$$

where α is chosen so that f_1 has a zero at $i\infty$. This is possible since $f_\lambda(i\infty) \neq 0$. Note that, since $k = mk_\lambda$, $f_1 \in M(\lambda, k, 1)$. If $f_1 \not\equiv 0$, let

$$f_2(\tau) = f_1(\tau)/f_\infty(\tau).$$

Both f_1 and f_∞ are analytic on $\mathcal{H}$, and the only zero of f_∞ on $\mathcal{H} \cup \{i\infty\}$, i.e., $i\infty$, is cancelled by the zero of f_1 at $i\infty$. Thus, f_2 is analytic on $\mathcal{H}$ and well defined at $i\infty$, that is, f_2 has a finite vertical limit at $i\infty$. Now,

$$f_2(-1/\tau) = \frac{(\tau/i)^k f_1(\tau)}{(\tau/i)^{4q/(q-2)} f_\infty(\tau)} = (\tau/i)^{k-4q/(q-2)} f_2(\tau).$$

Hence, $f_2 \in M(\lambda, k - k_\infty, 1)$, where $k_\infty = 4q/(q-2)$. Thus,

$$f(\tau) = f_1(\tau) + \alpha f_\lambda^m(\tau) = f_2(\tau) f_\infty(\tau) + \alpha f_\lambda^m(\tau). \tag{5.20}$$

Note that $k - k_\infty = mk_\lambda - qk_\lambda = (m-q)k_\lambda$. We now repeat with f_2 the argument that we made with f. We obtain a function f_4, say, such that $f_4 \in M(\lambda, (m-2q)k_\lambda, 1)$. Furthermore,

$$f_4(\tau) = f_3(\tau)/f_\infty(\tau),$$

where

$$f_3(\tau) = f_2(\tau) - \beta f_\lambda^{m-q}(\tau),$$

for a suitable constant β. Thus, we can write

$$f(\tau) = \{f_4(\tau) f_\infty(\tau) + \beta f_\lambda^{m-q}(\tau)\} f_\infty(\tau) + \alpha f_\lambda^m(\tau). \tag{5.21}$$

We continue this process until we get a function f^* with weight $m^* k_\lambda \leq qk_\lambda$, that is, $0 < m^* \leq q$. For with each application of the procedure we decrease the weight by qk_λ. Now consider

$$f^{**} = f^* - \alpha^* f_\lambda^{m^*},$$

where α^* is chosen so that $f^{**}(i\infty) = 0$. Clearly, $f^{**} \in M(\lambda, m^* k_\lambda, 1)$. From (5.16),

$$N + n_\infty + \frac{1}{2} n_i + \frac{n_\lambda}{q} = \frac{m^* k_\lambda (q-2)}{4q} = \frac{m^* k_\lambda}{k_\lambda q} = \frac{m^*}{q}.$$

Thus, f^{**} has m^*/q zeros. Since f^{**} has a zero at $i\infty$, we have a contradiction if $m^* < q$, unless $f^{**} \equiv 0$, i.e.,

$$f^* = \alpha^* f_\lambda^{m*}.$$

Suppose that $m^* = q$. Since f^{**} has a zero at $i\infty$, then f^{**} can have no other zeros. We also have both f^{**} and $f_\infty \in M(\lambda, qk_\lambda, 1)$, for $qk_\lambda = k_\infty$. From our remarks after the statement of Theorem 5.5, f^{**} is a constant multiple of f_∞. In summary, either

$$f^* = \alpha^* f_\lambda^{m^*}$$

or

$$f^* = \alpha^* f_\lambda^q + \beta^* f_\infty.$$

Hence, upon repeated substitution, we find eventually that f is of the form

$$f = \sum_{0 \le \nu \le [m/q]} b_\nu f_\infty^\nu f_\lambda^{m-q\nu}, \tag{5.22}$$

for some constants b_ν, $0 \le \nu \le [m/q]$. We have shown that every $f \in M(2\cos\pi/q, k, 1)$ can be written as a linear combination of the $[m/q]+1$ functions in the set

$$\mathcal{S} := \{f_\infty^\nu f_\lambda^{m-q\nu} : 0 \le \nu \le [m/q]\}.$$

On the other hand, it is clear that each of the functions in $\mathcal{S}$ lies in $M(2\cos\pi/q, k, 1)$.

Thus to complete the proof in the case $\gamma = 1$, it suffices to prove that the functions in $\mathcal{S}$ are linearly independent. Suppose, then, that there exist complex constants a_ν such that

$$\sum_{0 \le \nu \le [m/q]} a_\nu f_\infty^\nu f_\lambda^{m-q\nu} = 0,$$

or

$$\sum_{0 \le \nu \le [m/q]} a_\nu (f_\infty / f_\lambda^q)^\nu = 0 \qquad (\tau \ne \tau_\lambda).$$

By noting that $(f_\infty/f_\lambda^q)^\nu$ has a zero of order exactly ν at $i\infty$, we find that $a_\nu = 0$ for $0 \le \nu \le [m/q]$. It follows that the functions in $\mathcal{S}$ are linearly independent. Thus we have derived (5.19) in the case $\gamma = 1$.

Consider the case $\gamma = -1$. Let $f \in M(\lambda, k, -1)$, with $f \not\equiv 0$. Then n_i is odd and

$$f/f_i \in M(\lambda, k - 2q/(q-2), 1). \tag{5.23}$$

By the earlier part of the proof of the theorem (specifically (5.18) for the case $\gamma = 1$), if $k - 2q/(q-2) > 0$, then

$$k - \frac{2q}{q-2} = \frac{4m}{q-2} = mk_\lambda, \tag{5.24}$$

where m is a positive integer. It follows that

$$k = \frac{4m}{q-2} + \frac{2q}{q-2} = \frac{4(m+1)}{q-2} + 2,$$

and so (5.18) holds in this case.

Observe further that since $g \in M(\lambda, k - 2q/(q-2), 1)$, we conclude that $f = f_i g \in M(\lambda, k, -1)$. This, together with (5.23), implies that

$$\dim M(\lambda, k, -1) = \dim M(\lambda, k - 2q/(q-2), 1),$$

which, in turn, equals

$$1 + \left[\frac{m}{q}\right] = 1 + \left[\frac{m + 1 - 2/2}{q}\right],$$

by (5.19) for the case $\gamma = 1$. The theorem then follows for the case $\gamma = -1$, provided that $k > 2q/(q-2)$.

Suppose now that $k \leq 2q/(q-2)$, so that $k - 2q/(q-2)$ (the weight of f/f_i) is ≤ 0. If $k - 2q/(q-2) < 0$, an application of (5.16) to f/f_i leads to a contradiction unless $f/f_i \equiv 0$. It follows that $f \equiv 0$, and so there is nothing to prove. If $k - 2q/(q-2) = 0$, then $k = 2q/(q-2) = 4/(q-2) + 2$, and so (5.18) holds with $m = 1$. Furthermore, in Remark 5.3 immediately preceding the proof of this theorem, we established that

$$\dim M(\lambda, k - 2q/(q-2), 1) = 1,$$

so that

$$\dim M(\lambda, k, -1) = 1 = 1 + \left[\frac{2}{q}\right] = 1 + \left[\frac{1 + 2/2}{q}\right].$$

This, then, proves (5.19) for the case $k - 2q/(q-2) = 0$ and $\gamma = -1$ and, with it, Theorem 5.6 in its entirety. □

Remark 5.4. Observe that $f_i^2 \in M(\lambda, 4q/(q-2), 1)$. In the notation of (5.22) of Theorem 5.6, $m = q$. Since then $[m/q] = 1$, we see that, from (5.22),

$$f_i^2 = \alpha f_\lambda^q + \beta f_\infty,$$

for some constants α and β. By comparing the zeros, we find that both α and β are nonzero. In particular, f_∞ is a polynomial in f_i and f_λ. Hence, if $f \in M(\lambda, k, \gamma)$, it follows from (5.22) and (5.23) that f may be written as a polynomial in f_i and f_λ.

Theorem 5.6 gives us the precise dimension of $M(\lambda, k, \gamma)$. However, since we are interested in the number of Dirichlet series of signature (λ, k, γ), we wish to calculate $\dim M_0(\lambda, k, \gamma)$.

Remark 5.5. 1. When $\lambda = 1$, that is, $q = 3$ (the case of the modular group), Theorem 5.6 implies that $k \in 2\mathbb{Z}$.

2. It is worth noting that the restriction (5.18) on admissible weights k arises from condition (i) of Definition 2.2. Specifically, it is the *pure periodicity* of $f \in M(\lambda, k, \gamma)$ ($f(\tau + \lambda) = f(\tau)$ follows from (i)) that is the basis for (5.18).

For the remainder of this chapter we assume that $\lambda \in T_A$, so $\lambda = 2\cos(\pi/q)$, with $q \in \mathbb{Z}$ and $q \geq 3$.

Definition 5.2. If $f \in M(\lambda, k, \gamma)$ and $f(i\infty) = 0$, then we call f a *cusp form of weight* k *and multiplier* γ *with respect to* $G(\lambda)$. We denote by $C(\lambda, k, \gamma)$ the vector space of all cusp forms of weight k and multiplier γ with respect to $G(\lambda)$.

Observe that

$$\dim M(\lambda, k, \gamma) - \dim C(\lambda, k, \gamma) \leq 1. \tag{5.25}$$

Indeed, for any two elements in $M(\lambda, k, \gamma)$ not vanishing at $i\infty$, we may form a linear combination vanishing at $i\infty$. Thus, in a basis for $M(\lambda, k, \gamma)$ we may replace the aforementioned two elements not vanishing at $i\infty$ by one of the two elements and the constructed cusp form.

Theorem 5.7. *Let* k *be an arbitrary real number, and, as usual, let* $\tau = x + iy \in \mathcal{H}$.

(a) *For* $f \in C(\lambda, k, \gamma)$,

$$f(\tau) = O(y^{-k/2}),$$

uniformly in x, *as* $y \to 0+$.

(b) *For* $f \in M(\lambda, k, \gamma)$,

$$f(\tau) = \begin{cases} O(y^{-k}), & \text{if } k \geq 0, \\ O(y^{-k/2}), & \text{if } k < 0, \end{cases}$$

uniformly in x, *as* $y \to 0+$.

Proof of (a). Let

$$F(\tau) = F(x + iy) = |y^{k/2} f(\tau)|.$$

Let $V = (a\tau + b)/(c\tau + d) \in G(\lambda)$. Then, $\mathrm{Im}(V\tau) = y/|c\tau + d|^2$, and so

$$\begin{aligned} F(V\tau) &= |y^{k/2}(c\tau + d)^{-k} f(V\tau)| \\ &= |y^{k/2}(c\tau + d)^{-k}(c\tau + d)^k f(\tau)| \\ &= |y^{k/2} f(\tau)| = F(\tau). \end{aligned}$$

Hence, F is invariant under transformations in $G(\lambda)$. Now, since $f(i\infty) = 0$, $F(x + iy)$ tends to 0 as y tends to $+\infty$. Furthermore, $f(\tau)$ is continuous on the closed set $\{\tau : |x| \leq \lambda/2, |\tau| \geq 1\}$. These two facts imply that $F(\tau)$ is bounded on $\overline{B(\lambda)}$. Since $B(\lambda)$ is a fundamental region and F is invariant under transformations of $G(\lambda)$, for all $\tau \in \mathcal{H}$,

$$F(\tau) = |y^{k/2} f(\tau)| \leq A,$$

from which the result follows. □

Proof of (b). The idea of the proof is the same as in (a), but the details are somewhat more involved. Again let

$$F(\tau) = F(x + iy) = y^{k/2}|f(\tau)|;$$

as in the proof of (a), $F(V\tau) = F(\tau)$ for all $V \in G(\lambda)$. By Definition 2.2, $f(\tau)$ has an expansion at $i\infty$ of the form

$$f(\tau) = \sum_{n=0}^{\infty} a_n e^{2\pi i n\tau/\lambda}.$$

Thus since $\lambda < 2$ (so that $\overline{B(\lambda)}$ lies in the strip $\{\tau = x + iy : |x| \leq \frac{1}{2}, y \geq \frac{1}{2}\sqrt{4 - \lambda^2}\}$), it follows that

$$F(\tau) \leq K y^{k/2}, \qquad K > 0, \qquad \tau \in \overline{B(\lambda)} \cap \mathcal{H}. \tag{5.26}$$

The next step is to apply (5.26) and the invariance of F under $G(\lambda)$ to derive an inequality for $F(\tau)$ valid in all of $\mathcal{H}$. Toward this end, we prove the following lemma.

Lemma 5.3. (i) *If* $V = \left(\begin{smallmatrix} * & * \\ 0 & * \end{smallmatrix}\right) \in G(\lambda)$, *then* $V = \pm S_\lambda^n$, *with* $n \in \mathbb{Z}$.
(ii) *Suppose* $V = \left(\begin{smallmatrix} * & * \\ c & * \end{smallmatrix}\right) \in G(\lambda)$, *with* $c \neq 0$, *then* $c^2 \geq (1 + \lambda^2/4)^{-1}$.

Proof. We begin with the observation that, by the definition of $B(\lambda)$ (see Theorem 3.1), the set

$$\bigcup_{n \in \mathbb{Z}} S_\lambda^n\{\overline{B(\lambda)}\} \tag{5.27}$$

contains the closed upper half-plane $\{z : \operatorname{Im} z \geq 1\}$.

(i) Suppose $V = \left(\begin{smallmatrix} a & b \\ 0 & d \end{smallmatrix}\right) = \begin{pmatrix} a & b \\ 0 & 1/a \end{pmatrix} \in G(\lambda)$, so that $Vz = a^2 z + ab$. We claim that $V = \pm S_\lambda^n$, $n \in \mathbb{Z}$. Suppose not; then there exists $n_0 \in \mathbb{Z}$ such that $V(B(\lambda)) \cap S_\lambda^{n_0}(B(\lambda))$ is a nonempty open set. Since $V \neq \pm S_\lambda^{n_0}$, this contradicts condition (i) of Definition 3.3, which $B(\lambda)$ satisfies by Theorem 3.1.

(ii) Let $V = \left(\begin{smallmatrix} * & * \\ c & d \end{smallmatrix}\right) \in G(\lambda)$, with $c \neq 0$. Put

$$V' = VS_\lambda^n = \begin{pmatrix} * & * \\ c & d + n\lambda c \end{pmatrix},$$

with $n \in \mathbb{Z}$. We can choose $n \in \mathbb{Z}$ so that $d' = d + n\lambda c$ satisfies the inequality $|d'| \leq \lambda|c|/2$. On the other hand, since $c \neq 0$, by (5.27) and condition (i) of Definition 3.3, $\operatorname{Im}(V\tau) \leq 1$ for $\tau \in \overline{B(\lambda)}$. Thus,

$$\frac{\operatorname{Im} \tau}{|c\tau + d'|^2} = \operatorname{Im}(V'\tau) \leq 1, \quad \text{or} \quad |c\tau + d'|^2 \geq y,$$

for $\tau \in \overline{B(\lambda)}$. Applying this inequality with $\tau = i \in \overline{B(\lambda)}$, we find that

$$c^2(1 + \lambda^2/4) = c^2 + (\lambda^2/4)c^2 \geq c^2 + d'^2 = |ci + d'|^2 \geq 1;$$

that is to say, $c^2 \geq (1 + \lambda^2/4)^{-1}$, as claimed. □

Remark 5.6. Since $B(\lambda)$ is actually the same as the Ford Fundamental Region

$$B^*(\lambda) = \left\{\tau \in H : |\operatorname{Re} \tau| < \lambda/2,\ |c\tau + d| > 1\ \forall\, V = \begin{pmatrix} * & * \\ c & * \end{pmatrix} \in G(\lambda),\ c \neq 0\right\}$$

(see [73, p. 139] or [74, pp. 57–58]), the inequality $\dfrac{\operatorname{Im} \tau}{|c\tau + d'|^2} \leq 1$ that we derived in the proof of Lemma 5.3(ii) can be supplanted by the inequality $\dfrac{\operatorname{Im} \tau}{|c\tau + d'|^2} \leq \operatorname{Im}\, \tau$, which is sometimes stronger. This supplemental inequality could be used to prove Lemma 5.3(ii), but the proof that $B^*(\lambda) = B(\lambda)$ would require an extensive digression.

We continue the proof of Theorem 5.7(b). Given $\tau = x + iy$ in $\mathcal{H}$, there exist $z \in \overline{B(\lambda)} \cap \mathcal{H}$ and $V = \left(\begin{smallmatrix} a & b \\ c & d \end{smallmatrix}\right) \in G(\lambda)$ such that $\tau = Vz$, by property

(ii) of Definition 3.3. Then

$$\begin{aligned}|f(\tau)| &= y^{-k/2}F(\tau) = y^{-k/2}F(Vz) = y^{-k/2}F(z)\\ &\leq Ky^{-k/2}(\operatorname{Im} z)^{k/2},\end{aligned}$$

by (5.26).

Suppose that $k \geq 0$. We have

$$\operatorname{Im} z = \operatorname{Im}(V^{-1}\tau) = \frac{\operatorname{Im}\tau}{|-c\tau + a|^2} = \frac{y}{|-c\tau + a|^2}.$$

If $c = 0$, then by Lemma 5.3(i), it follows that $\operatorname{Im} z = \operatorname{Im}\tau$. If $c \neq 0$, then by Lemma 5.3(ii),

$$|-c\tau + a|^2 = c^2y^2 + (a - cx)^2 \geq c^2y^2 \geq \left(1 + \frac{1}{4}\lambda^2\right)^{-1} y^2.$$

In this case it follows that $\operatorname{Im} z \leq \left(1 + \frac{1}{4}\lambda^2\right) y^{-1}$. In either case, then, $\operatorname{Im} z \leq y + \left(1 + \frac{1}{4}\lambda^2\right) y^{-1}$, and we conclude that

$$|f(\tau)| \leq Ky^{-k/2}\left\{y + \left(1 + \frac{1}{4}\lambda^2\right) y^{-1}\right\}^{k/2} \leq K_1 y^{-k}, \tag{5.28}$$

for all $y \geq 1$, say. This completes the proof for the case $k \geq 0$.

Suppose that $k < 0$. We return to the inequality

$$|f(\tau)| = y^{-k/2}F(\tau) = y^{-k/2}F(Vz) = y^{-k/2}F(z) \leq Ky^{-k/2}(\operatorname{Im} z)^{k/2}.$$

Since $k < 0$, we require a lower bound for $\operatorname{Im} z$. But, since $z \in \overline{B(\lambda)}$, it follows that $\operatorname{Im} z \geq \sqrt{1 - \frac{1}{4}\lambda^2}$, so that $|f(\tau)| \leq K_2 y^{-k/2}$. This completes the proof of Theorem 5.7. □

Corollary 5.3. (a) *For arbitrary real* k,

$$M(\lambda, k, \gamma) = M_0(\lambda, k, \gamma).$$

In particular, $C(\lambda, k, \gamma) \subset M_0(\lambda, k, \gamma)$.

(b) *If* $k < 0$, *then*

$$M(\lambda, k, \gamma) = M_0(\lambda, k, \gamma) = \{0\}.$$

In particular, $C(\lambda, k, \gamma) = \{0\}$.

Proof. (a) Since $M_0(\lambda, k, \gamma) \subset M(\lambda, k, \gamma)$ by definition, it suffices to prove the reverse inequality. But this follows from Theorem 5.7(b) and Lemma 2.2.

(b) Let $f(\tau) \in M(\lambda, k, \gamma)$, and consider the expansion at $i\infty$ given in Definition 2.2, namely,

$$f(\tau) = \sum_{n=0}^{\infty} a_n e^{2\pi i n\tau/\lambda}, \quad \tau \in \mathcal{H}.$$

By interchanging the order of summation and integration, we find that

$$a_n = \frac{1}{\lambda} \int_{\tau_0}^{\tau_0+\lambda} f(\zeta) e^{-2\pi i n\zeta/\lambda} d\zeta,$$

where the integration is along a horizontal path, and where τ_0 is an arbitrary point in $\mathcal{H}$. Then, Theorem 5.7(b) implies that

$$|a_n| \leq K y_0^{-k/2} e^{2\pi n y_0/\lambda}, \qquad y_0 = \operatorname{Im} \tau_0 > 0.$$

Since $-k/2 > 0$, we let $y_0 \to 0+$ to deduce that $a_n = 0$ for all $n \geq 0$. This implies that $f(\tau) \equiv 0$. □

Remark 5.7. 1. If we invoke the method of proof of Corollary 5.3(b) in the case $k = 0$, we obtain $|a_n| \leq K$, a result insufficient to prove that f is, in fact, constant. The latter result does hold, however (see [63]).

2. Hecke [46, p. 222] applied the method of proof of Corollary 5.3(b) and the estimate in Theorem 5.7(a) to obtain $a_n = O(n^{k/2})$ for the coefficients a_n of cusp forms. The same method, together with Theorem 5.7(b), yields $a_n = O(n^k)$ for the coefficients of entire forms ($k \geq 0$).

3. Note that Corollary 5.3(b) follows directly from Lemma 5.1.

Theorem 5.8. *We have*

$$\dim C(\lambda, k, \gamma) = \left[\frac{m + (\gamma - 1)/2}{q}\right],$$

where $k = 4m/(q-2) + 1 - \gamma$.

Proof. By Theorem 5.6,

$$\dim M(\gamma, k, \gamma) = 1 + \left[\frac{m + (\gamma - 1)/2}{q}\right].$$

On the other hand, in the proof of Theorem 5.5 we constructed automorphic forms f_λ, f_i with the properties $f_\lambda \in M(\lambda, 4/(q-2), 1)$, $f_\lambda(i\infty) \neq 0$; $f_i \in$

$M(\lambda, 2q/(q-2), -1)$, $f_i(i\infty) \neq 0$. Furthermore, by Theorem 5.6, the only admissible values for k have the form (5.18). But $f_\lambda^m \in M(\lambda, 4m/(q-2), 1)$ with $f_\lambda^m(i\infty) \neq 0$ and $f_i f_\lambda^{m-1} \in M(\lambda, 4m/(q-2)+2, -1)$ with $f_i(i\infty) f_\lambda^{m-1}(i\infty) \neq 0$. Thus, by (5.25), $\dim M(\lambda, k, \gamma) = \dim M_0(\lambda, k, \gamma) = \dim C(\lambda, k, \gamma) + 1$, and so Theorem 5.8 follows from (5.19). □

Corollary 5.4. *The dimension of the space of entire Dirichlet series of signature* (λ, k, γ) *is*

$$\left[\frac{m + (\gamma - 1)/2}{q}\right].$$

We now give some examples of functions belonging to $M_0(1, k, \gamma)$. First we prove a lemma.

Lemma 5.4. *Let* $k > 2$. *Then*

$$\sum_{m,n=-\infty}^{\infty}{}' |m\tau + n|^{-k}$$

converges for every $\tau \in \mathcal{H}$. *Furthermore, the convergence is uniform on compact subsets of* $\mathcal{H}$. *Here, the dash* $'$ *indicates that the term with* $m = n = 0$ *is omitted from the summation.*

Proof. Fix $\tau \in \mathcal{H}$ and let

$$S_r = \{\pm r\tau + n; n\tau \pm r : -r \leq n \leq r\},$$

where $r \geq 1$ is an integer. Note that the set S_r contains $8r$ points. Let $h = h(\tau)$ denote the distance of S_1 from the origin. Then hr is the distance of S_r from the origin. Thus, if $m\tau + n \in S_r$, then $|m\tau + n| \geq hr$. Hence,

$$\sum_{m\tau+n \in S_r} |m\tau + n|^{-k} \leq 8r(rh)^{-k}$$

and

$$\sum_{m,n=-\infty}^{\infty}{}' |m\tau + n|^{-k} = \sum_{r=1}^{\infty} \sum_{m\tau+n \in S_r} |m\tau + n|^{-k} \leq 8h^{-k} \sum r^{-k+1} < \infty,$$

if $k > 2$. If τ belongs to some compact set S in $\mathcal{H}$, then $\{h(\tau)\}^{-k} \leq A$. By Weierstrass's M-test the convergence is uniform on compact subsets of $\mathcal{H}$. □

Definition 5.3. For $k > 2$ and $\tau \in \mathcal{H}$, define the Eisenstein series $G_k(\tau)$ by

$$G_k(\tau) = \sum_{m,n=-\infty}^{\infty}{}' (m\tau + n)^{-k}.$$

Theorem 5.9. *If $k > 2$ is an even integer, $G_k(\tau) \in M_0(1, k, (-1)^{k/2})$. Furthermore, for $\tau \in \mathcal{H}$,*

$$G_k(\tau) = 2\zeta(k) + \frac{2(2\pi i)^k}{\Gamma(k)} \sum \sigma_{k-1}(n) e^{2\pi i \tau n}, \tag{5.29}$$

where

$$\sigma_a(n) = \sum_{\substack{d|n \\ d>0}} d^a.$$

Proof. Since $k > 2$, $G_k(\tau)$ is analytic on $\mathcal{H}$ by Lemma 5.4. Now,

$$\begin{aligned} G_k(-1/\tau) &= \tau^k \sum_{m,n=-\infty}^{\infty}{}' (-m + n\tau)^{-k} \\ &= \tau^k G_k(\tau) = (-1)^{k/2} (\tau/i)^k G_k(\tau). \end{aligned}$$

Thus, property (ii) of Definition 2.2 is valid. If we prove (5.29), (i) and (iii) of Definition 2.2 will follow since it is well known that $\sigma_{k-1}(n)$ satisfies (iii).

Since k is even, we have, upon separating the terms with $m = 0$,

$$G_k(\tau) = 2\zeta(k) + 2 \sum_{m=1}^{\infty} \sum_{n=-\infty}^{\infty} (m\tau + n)^{-k}. \tag{5.30}$$

It is well known [1, p. 187] that for $\tau \in \mathcal{H}$,

$$\frac{\pi^2}{\sin^2 \pi\tau} = \sum_{n=-\infty}^{\infty} (n + \tau)^{-2}.$$

Thus,

$$\begin{aligned} \sum_{n=-\infty}^{\infty} (n + \tau)^{-2} = \frac{(2\pi i)^2}{(e^{\pi i \tau} - e^{-\pi i \tau})^2} &= \frac{(2\pi i)^2 e^{2\pi i \tau}}{(1 - e^{2\pi i \tau})^2} \\ &= (2\pi i)^2 \sum n e^{2\pi i n \tau}. \end{aligned}$$

Differentiate both sides, above, $k-2$ times with respect to τ to obtain

$$(k-1)!\sum_{n=-\infty}^{\infty}(n+\tau)^{-k}=(2\pi i)^k\sum n^{k-1}e^{2\pi in\tau}. \tag{5.31}$$

Replacing τ by $m\tau$ and summing on m, we arrive at

$$\begin{aligned}\sum_{m=1}^{\infty}\sum_{n=-\infty}^{\infty}(m\tau+n)^{-k}&=\frac{(2\pi i)^k}{\Gamma(k)}\sum\sum n^{k-1}e^{2\pi imn\tau}\\&=\frac{(2\pi i)^k}{\Gamma(k)}\sum\sigma_{k-1}(n)e^{2\pi in\tau}.\end{aligned} \tag{5.32}$$

Combining (5.32) with (5.30), we arrive at (5.29), and so the proof is complete. □

Remark 5.8. The formula (5.31) is a special case of the Lipschitz summation formula. (See [60, p. 65].)

Corollary 5.5. *If* $k>2$ *is even,* $\zeta(s)\zeta(s-k+1)$ *has signature* $(1,k,(-1)^{k/2})$.

Proof. Since

$$\zeta(s)\zeta(s-k+1)=\sum\sigma_{k-1}(n)n^{-s},$$

the result follows from Theorem 5.9. □

Note that $f_1\in M_0(1,4,1)$ and $f_i\in M_0(1,6,-1)$. Here, $f_1=f_\lambda$ (in the notation of Theorem 5.5) when $\lambda=1$, and is not to be confused with either of the functions with the same designation in the proofs of Theorems 5.5 and 5.6. Since $\dim M_0(1,4,1)=\dim M_0(1,6,-1)=1$, it follows that G_4 is a multiple of f_1 and G_6 is a multiple of f_i. In particular, the only zero of $G_4(\tau)$ on $\overline{B(1)}$ (up to equivalence) is a simple zero at $\exp(2\pi i/3)$.

Definition 5.4. For even $k>2$, the *normalized Eisenstein series* $E_k(\tau)$ is defined by

$$E_k(\tau)=\frac{G_k(\tau)}{2\zeta(k)}=1+(-1)^{k/2}A_k\sum\sigma_{k-1}(n)e^{2\pi in\tau},$$

where

$$A_k:=\frac{(2\pi)^k}{\Gamma(k)\zeta(k)}.$$

Theorem 5.10. *The numbers* A_k *are rational, when* $k\geq 2$ *is even.*

Proof. We begin with the familiar formula [105, p. 114]

$$\sin z = z \prod_{n=1}^{\infty} \left(1 - \frac{z^2}{n^2\pi^2}\right).$$

It follows that for $|z| < \pi$,

$$\begin{aligned} z \cot z &= z \frac{d}{dz} \log(\sin z) \\ &= 1 - 2z^2 \sum \frac{1}{n^2\pi^2(1 - z^2/n^2\pi^2)} \\ &= 1 - 2 \sum_{n=1}^{\infty} \sum_{m=0}^{\infty} (z^2/n^2\pi^2)^{m+1} \\ &= 1 - 2 \sum_{m=1}^{\infty} (z/\pi)^{2m} \sum_{n=1}^{\infty} n^{-2m} \\ &= 1 - 2 \sum (z/\pi)^{2m} \zeta(2m). \end{aligned} \tag{5.33}$$

On the other hand, it is clear that the Maclaurin series coefficients of $z \cot z$ are rational. Hence, $\zeta(2m)/\pi^{2m}$ is rational, $m = 1, 2, \ldots$, and the result follows. □

Definition 5.5. If $k \geq 2$ is even, $B_k = 2k/A_k$ is the kth *Bernoulli number.*

We now construct some cusp forms for the full modular group. First, note that, by Theorem 5.8,

$$\dim C(1, k, \gamma) = \begin{cases} \left[\frac{1}{12}k\right], & \text{if } \gamma = 1, \\ \left[\frac{1}{12}k - \frac{1}{6}\right], & \text{if } \gamma = -1, \end{cases}$$

or, if $k \geq 4$,

$$\dim C(1, k, \gamma) = \begin{cases} \left[\frac{1}{12}k\right], & \text{if } k \not\equiv 2 \pmod{12}, \\ \left[\frac{1}{12}k\right] - 1, & \text{if } k \equiv 2 \pmod{12}. \end{cases}$$

Definition 5.6. For $\tau \in \mathcal{H}$, the classical *discriminant function* $\Delta(\tau)$ is defined by

$$\Delta(\tau) = \frac{E_4(\tau)^3 - E_6(\tau)^2}{1728}.$$

Since the Fourier expansions of E_4 and E_6 both begin with 1, it is clear that $\Delta(\tau) \in C(1, 12, 1)$. Note that 12 is the least value of k for which there exists a nontrivial cusp form. Now, $A_4 = 240$ and $A_6 = 504$. This can be seen, for example, by a direct calculation of the Maclaurin series for $z \cot z$ and the use of (5.33). Thus,

$$\begin{aligned}\Delta(\tau) &= \frac{1}{1728}\left\{\left(1 + 240\sum \sigma_3(n)e^{2\pi i n\tau}\right)^3 - \left(1 - 504\sum \sigma_5(n)e^{2\pi i n\tau}\right)^2\right\} \\ &= \sum a_n e^{2\pi i n\tau}, \qquad (5.34)\end{aligned}$$

say. Since $\sigma_3(1) = \sigma_5(1) = 1$,

$$a_1 = \frac{3 \cdot 240 + 2 \cdot 504}{1728} = 1.$$

In fact, as we shall see below, a_n is integral for all n.

Definition 5.7. The classical *Dedekind eta function* $\eta(\tau)$ is defined for $\tau \in \mathcal{H}$ by

$$\eta(\tau) = e^{\pi i\tau/12}\prod_{n=1}^{\infty}(1 - e^{2\pi i n\tau}). \qquad (5.35)$$

It is well known [60, p. 43] that $\eta(\tau)$ is a modular form (in a more general sense than we have defined here) of weight $\frac{1}{2}$ on the full modular group. Indeed, the behavior of $\eta(\tau)$ under $\Gamma(1) = G(1)$ is analogous to that of $\theta(\tau)$ under the subgroup (of index 3) $\Gamma_\theta = G(2)$. (For example, (1.1) holds with $\theta(\tau)$ replaced by $\eta(\tau)$.) Since the translation property $\eta(\tau + 1) = e^{\pi i/12}\eta(\tau)$ follows directly from the definition (5.35) and since $G(1) = \Gamma(1) =< S, T >$, it follows that $\eta(\tau)$ transforms into itself under transformations in $G(1)$. (See [60, pp. 43–45, proof of Theorem 10] and the argument in Chapter 7 here, immediately preceding Remark 7.7.) Thus,

$$\eta(\tau)^{24} = e^{2\pi i\tau}\prod_{n=1}^{\infty}(1 - e^{2\pi i n\tau})^{24} \in C(1, 12, 1).$$

Since $\dim C(1, 12, 1) = 1$ and both $\eta(\tau)^{24}$ and $\Delta(\tau)$ have leading coefficient 1, $\eta(\tau)^{24} = \Delta(\tau)$. It is also clear that a_n is integral for every n. The coefficients a_n are usually denoted by $\tau(n)$, Ramanujan's arithmetical function. For an excellent introduction to the properties of $\tau(n)$ see [43, Chapter 10]. Further informative expository papers have been written by F. van der Blij [108], M. R. Murty [84], V. K. Murty [85], R. A. Rankin [93], and H. P. F. Swinnerton-Dyer [102], with the papers by van der Blij,

M. R. Murty, and Rankin containing lengthy lists of references. In 1916 Ramanujan conjectured that $\tau(n)$ is multiplicative, and this was proved by L. J. Mordell [80] in 1917. (It is of historical interest that in [80] Mordell invented the "Hecke operators" in order to prove the multiplicativity of $\tau(n)$.) Mordell also showed that for $\sigma = \operatorname{Re} s$ sufficiently large,

$$\sum \tau(n) n^{-s} = \prod_p (1 - \tau(p) p^{-s} + p^{11-2s})^{-1},$$

where the product is taken over all primes. The product actually converges absolutely for $\sigma > 13/2$. A second deep conjecture of Ramanujan,

$$|\tau(p)| \leq 2p^{11/2},$$

where p is any prime, resisted all efforts to prove it for nearly sixty years, until finally P. Deligne succeeded in 1974 [28]. A third well-known problem concerning $\tau(n)$ is Lehmer's question whether $\tau(n)$ is ever 0 [72]. There is a good deal of numerical evidence to support the proposition that $\tau(n) \neq 0$ for all $n \in \mathbb{Z}^+$, but the problem appears very difficult. (The nonvanishing of $\tau(n)$ is often referred to as "Lehmer's conjecture," even though he never conjectured it.)

We close this chapter by indicating how one can construct Dirichlet series of signature (λ, k, γ) from those of signature $(1, k, \gamma)$ for certain λ. First, if λ is a positive integer the problem is trivial, for in the notation of Theorem 2.1,

$$\Phi(s) = (2\pi)^{-s} \Gamma(s) \varphi(s), \qquad \varphi(s) = \sum a_n n^{-s},$$

may be rewritten as

$$\Phi^*(s) = \left(\frac{2\pi}{\lambda}\right)^{-s} \Gamma(s) \varphi^*(s), \qquad \varphi^*(s) = \sum a_n (\lambda n)^{-s},$$

and $\varphi^*(s)$ is again a Dirichlet series, since $\lambda \in Z^+$. On the other hand, $\varphi(s)$ has signature $(1, k, \gamma)$ and $\Phi^*(s) = \Phi(s)$, and so $\Phi^*(k - s) = \gamma \Phi^*(s)$. It follows that $\varphi^*(s)$ has signature (λ, k, γ).

We now construct functions in the space $M_0(\lambda^{\frac{1}{2}}, k, \gamma)$ from those in $M_0(1, k, \gamma)$, where λ is a positive integer. Let $f \in M_0(1, k, \gamma)$ and put $g(z) = f(z/\lambda)$, $z \in \mathcal{H}$. Then,

$$f(-\lambda/z) = \gamma(z/i\lambda)^k f(z/\lambda) = \gamma(z/i\lambda)^k g(z)$$

and

$$g(-\lambda/z) = f(-1/z) = \gamma(z/i)^k f(z).$$

If we put

$$H(z) = f(z) + \lambda^{-k/2} g(z),$$

it follows that

$$\begin{aligned} H(-\lambda/z) &= \gamma(z/i\lambda)^k g(z) + \gamma(z/i)^k \lambda^{-k/2} f(z) \\ &= \gamma(z/i)^k \lambda^{-k/2} H(z). \end{aligned}$$

Note also that $H(z+\lambda) = H(z)$.

Now put $z = \sqrt{\lambda}\tau$, $\tau \in \mathcal{H}$. Define

$$F(\tau) = H(z) = H\left(\sqrt{\lambda}\tau\right).$$

Then,

$$F\left(\tau + \sqrt{\lambda}\right) = H\left(\sqrt{\lambda}\tau + \lambda\right) = H\left(\sqrt{\lambda}\tau\right) = F(\tau),$$

and

$$\begin{aligned} F(-1/\tau) = H\left(-\sqrt{\lambda}/\tau\right) = H(-\lambda/z) &= \gamma(z/i)^k \lambda^{-k/2} H(z) \\ &= \gamma(\tau/i)^k F(\tau). \end{aligned}$$

Hence, $F \in M_0\left(\sqrt{\lambda}, k, \gamma\right)$.

Chapter 6

The case $\lambda = 2$

The group $G(2)$ is, in fact, a subgroup of index 3 of the modular group $G(1)$. (See [60, p. 9].) Recall that, for $\lambda < 2$ and $\lambda \in T_A$, we showed that the function $g(\tau)$ in Theorem 5.5 has a zero of order q at τ_λ in the standard planar local variable $\tau - \tau_\lambda$. From $g(\tau)$ we were able to construct a basis for $M(\lambda, k, \gamma)$. The behavior of a modular form with respect to $G(2)$ in a neighborhood of $\tau_\lambda = \tau_2 = -1$ will be more difficult to determine as $-1 \notin \mathcal{H}$. We would first like to establish a lemma analogous to Lemma 5.1. However, to do this, a study of functions in $M(2, k, \gamma)$ near $\tau = -1$ is necessary.

Our initial task is to find a local uniformizing variable near $\tau = -1$ for $f \in M(2, k, \gamma)$. It will be convenient to transform $\tau = -1$ to $i\infty$ by the modular transformation

$$z = z(\tau) = -1/(\tau + 1). \tag{6.1}$$

Note that -1 is fixed under the transformation

$$V(\tau) = -1/(\tau + 2) \tag{6.2}$$

in $G(2)$. Now,

$$z(V\tau) = \frac{-1}{-1/(\tau+2)+1} = -\frac{\tau+2}{\tau+1} = z(\tau) - 1. \tag{6.3}$$

Inverting (6.1), we find that $\tau = \tau(z) = -(z+1)/z$. Another elementary calculation like that in (6.3) gives $\tau(z-1) = -1/(\tau+2) = V(\tau)$. That is to say, the map $\tau \to V(\tau)$ is equivalent to the map $z \to z-1$. This observation can be applied to show that $t(\tau) = \exp\{2\pi i V(\tau)\} = \exp\{-2\pi i/(\tau+1)\}$ is a local variable at the cusp -1 with respect to $G(2)$.

In order to establish this, it suffices to show that there exists a subset $\mathcal{N}(-1,2) = \mathcal{N}$ in a punctured neighborhood U of -1 such that $t(\tau)$ has properties (i), (ii), and (iii) of Definition 4.1. Since our functions are defined only on $\mathcal{H} \cup Q^+ = \mathcal{H} \cup Q \cup \{i\infty\}$, we must first define "punctured neighborhoods" of -1 in such a way that they lie inside $\mathcal{H}$.

The idea for doing this is motivated by the definition we gave in Chapter 4 of a "punctured neighborhood" of $i\infty$. This is defined as any open half-plane of the form $\operatorname{Im} \tau > B$, with $B > 0$. Observe that $\tau \to z(\tau)$ maps -1 to $i\infty$ and the circle of radius $1/(2y_0)$, $y_0 > 0$, centered at $-1 + i/(2y_0)$ (thus tangent to the real line at $\tau = -1$) onto the horizontal line at height y_0. Furthermore, let D be the open disc bounded by the circle above; then $z(\tau)$ maps D onto the open half-plane $\operatorname{Im} \tau > y_0$ in a one-to-one fashion. This suggests the requisite definition.

Definition 6.1. A *punctured neighborhood of* -1 is any open disc $U = U(y_0)$ of the form

$$(x+1)^2 + \left(y - \frac{1}{2y_0}\right)^2 = \frac{1}{2y_0}, \tag{6.4}$$

with $\tau = x + iy$ and $y_0 > 0$.

We now define the required subset $\mathcal{N} = \mathcal{N}(-1,2) = \mathcal{N}(-1,2;y_0)$ to be the "triangle" in $\mathcal{H}$ bounded by the three circular arcs:

(a) the circle defined by (6.4),

(b) the unit circle $x^2 + y^2 = 1$,

(c) the circle $(x+2)^2 + y^2 = 1$.

Note that one of the vertices of the triangle lies at the point -1. A simple calculation shows that $\tau \to z(\tau)$ maps this triangle onto the open half-strip

$$\left\{ z \in \mathcal{H} : |\operatorname{Re} z| < \frac{1}{2}, \operatorname{Im} z > y_0 \right\} \tag{6.5}$$

in a one-to-one manner.

We have previously observed that $\tau \to V(\tau)$ is equivalent to $z \to z - 1$. Also, distinct points of the half-strip (6.5) are inequivalent with respect to the map $z \to z-1$. Thus, since by Theorem 3.1 $B(2)$ is a fundamental region for $G(2)$, it follows readily that for $y_0 > 0$ sufficiently large ($y_0 > 1$ will do), distinct points of $\mathcal{N}(-1,2;y_0)$ are inequivalent modulo $G(2)$. Furthermore

$$t = \exp\{2\pi i z(\tau)\} = \exp\{-2\pi i/(\tau+1)\}$$

maps $\mathcal{N}(-1, 2; y_0)$ one-to-one onto the punctured disc $D' : 0 < |t| < e^{-2\pi y_0}$, since this is the image of the half-strip (6.5) under $z \to \exp(2\pi i z)$. Thus, $t(\tau) = \exp\{-2\pi i/(\tau+1)\}$ is indeed a local uniformizing variable at -1 with respect to $G(2)$. We show that a modular form with respect to $G(2)$, of any weight, has an expansion at -1 in the variable $t(\tau)$. This is the content of the following result.

Theorem 6.1. (a) *Let $f \in M(2, k, \gamma)$. Then there exists a real number ρ, $0 \le \rho < 1$, such that for $\tau \in \mathcal{H}$,*

$$f(\tau) = \left(\frac{\tau+1}{i}\right)^{-k} \sum_{n=-\infty}^{\infty} a_n e^{2\pi i(n+\rho)z(\tau)}$$
$$= \left(\frac{\tau+1}{i}\right)^{-k} \sum_{n=-\infty}^{\infty} a_n e^{-2\pi i(n+\rho)/(\tau+1)}, \qquad (6.6)$$

where $\{a_n\}$ is a sequence of complex numbers.

(b) *If $f \in M_0(2, k, \gamma)$, then the expansion (6.6) has no terms with $n + \rho < 0$; that is, (6.6) takes the form*

$$f(\tau) = \left(\frac{\tau+1}{i}\right)^{-k} \sum_{n+\rho\ge 0} a_n e^{-2\pi i(n+\rho)/(\tau+1)}. \qquad (6.7)$$

Proof. (a) From (i) and (ii) of Definition 2.2 and (6.2),

$$f(V\tau) = \gamma\left(\frac{\tau+2}{i}\right)^{k} f(\tau+2) = \gamma\left(\frac{\tau+2}{i}\right)^{k} f(\tau). \qquad (6.8)$$

From (6.8) and the fact that

$$z(\tau)(\tau+2) = -(\tau+2)/(\tau+1) = z(\tau) - 1,$$

we find that

$$f(\tau)\left(\frac{z(\tau)}{i}\right)^{-k} = \gamma\left(\frac{\tau+2}{i}\right)^{-k} f(V\tau)\left(\frac{z(\tau)}{i}\right)^{-k}$$
$$= \epsilon\left(\frac{z(\tau)-1}{i}\right)^{-k} f(V\tau), \qquad (6.9)$$

for some number ϵ with $|\epsilon| = 1$. Write $\epsilon = \exp(2\pi i\rho)$, $0 \le \rho < 1$. Thus, from (6.9), we find that

$$f(\tau)\left(\frac{z(\tau)}{i}\right)^{-k} = e^{2\pi i\rho}\left(\frac{z(\tau)-1}{i}\right)^{-k} f(V\tau). \qquad (6.10)$$

But, as we have seen, the map $\tau \to V\tau$ is equivalent to the map $z \to z-1$. Thus, by (6.10), the function

$$H(z) = e^{-2\pi i\rho z}(z/i)^{-k}f(\tau) = e^{-2\pi i\rho z}(z/i)^{-k}f\left(-\frac{1}{z}-1\right)$$

is periodic in z, with period -1. Since $f(\tau)$ is holomorphic for τ in $\mathcal{H}$, $H(z)$ is holomorphic in z for $\operatorname{Im} z > 0$. Thus, $H(z)$ has a Laurent expansion in the local variable $t = \exp\{2\pi iz\} = \exp\{2\pi iz(\tau)\}$, namely,

$$H(z) = \sum_{n=-\infty}^{\infty} a_n e^{2\pi inz},$$

or

$$\begin{aligned} f(\tau) &= (z/i)^k \sum_{n=-\infty}^{\infty} a_n e^{2\pi i(n+\rho)z} \\ &= \left(\frac{\tau+1}{i}\right)^{-k} \sum_{n=-\infty}^{\infty} a_n e^{-2\pi i(n+\rho)/(\tau+1)}, \end{aligned}$$

for τ in $\mathcal{H}$.

(b) Now we assume that $f \in M_0(2,k,\gamma)$ and show that $a_n = 0$ if $n+\rho < 0$.

From the standard formula for Fourier coefficients,

$$a_n = \int_{z_0}^{z_0+1} f(\tau)\left(\frac{\tau+1}{i}\right)^k e^{-2\pi i(n+\rho)z}\,dz.$$

Let $z_0 = u_0 + iv_0$ and $z = u + iv_0$ with $u_0 \le u \le u_0 + 1$ and v_0 very large. Since, by Lemma 2.2,

$$f(x+iy) = O(y^{-c})$$

as $y > 0$ tends to 0, for some $c > 0$,

$$f(\tau) = f\left(-1-\frac{1}{z}\right) = f\left(-1-\frac{u-iv_0}{(u^2+v_0^2)}\right) = O(v_0^c),$$

as v_0 tends to ∞. Hence,

$$a_n = O(v_0^{c-k}e^{2\pi v_0(n+\rho)}),$$

as v_0 tends to ∞. Hence, if $n+\rho < 0$, we must necessarily have $a_n = 0$, and this proves (b). □

Definition 6.2. Let $M_1(2,k,\gamma)$ denote the subspace of $M(2,k,\gamma)$ satisfying condition (6.7).

Theorem 6.1 shows that $M_0(2,k,\gamma) \subseteq M_1(2,k,\gamma) \subseteq M(2,k,\gamma)$.

Definition 6.3. In (6.7), suppose that $a_n = 0$, $0 \leq n \leq m-1$, but $a_m \neq 0$. Then, we say that $f(\tau)$ *has a zero of order* $n_{-1} = m + \rho$ *at* $\tau = -1$.

Remark 6.1. In general, the order of the zero of f at $\tau = -1$ will be irrational. However, suppose that $f \in M(1, k, (-1)^{k/2})$; *a fortiori*, then, $f \in M_0(2, k, (-1)^{k/2})$. Since, in this case,

$$f\left(-\frac{1}{\tau+1}\right) = \gamma\left(\frac{\tau+1}{i}\right)^k f(\tau+1) = \gamma\left(\frac{\tau+1}{i}\right)^k f(\tau),$$

a straightforward calculation shows that $n_{-1} = n_\infty$. Hence, since n_∞ is an integer, n_{-1} is an integer as well. (Recall that n_∞, the order of the zero of f at $i\infty$, is defined to be the smallest exponent occurring in the expansion (i) of Definition 2.2.)

Lemma 6.1. *Let* N, n_i, *and* n_∞ *be defined as in the beginning of Chapter 5. Let* n_{-1} *be given as above. Then if* $f \in M_1(2,k,\gamma)$ *and* $f \not\equiv 0$,

$$N + n_\infty + \frac{1}{2}n_i + n_{-1} = \frac{k}{4} \tag{6.11}$$

and

$$\gamma = (-1)^{n_i}.$$

Proof. First, from (6.7),

$$e^{-2\pi i\rho z} f(\tau)\left(\frac{\tau+1}{i}\right)^k = \sum_{n+\rho\geq 0} a_n t^n,$$

where $t = \exp(2\pi iz)$. It follows that $f(\tau)$ has only a finite number of zeros in the intersection of $B(2)$ with any neighborhood of $\tau = -1$. For if there were an infinite number of zeros, then

$$\sum_{n+\rho\geq 0} a_n t^n \tag{6.12}$$

would have an infinite number of zeros in a neighborhood of $t = 0$. This would imply that (6.12) is identically zero, and so $f(\tau) \equiv 0$, contrary to our assumption.

We can then proceed as in the proof of Lemma 5.1 to derive (6.11). The contour of integration is the same as in the proof of Lemma 5.1, except that now $\lambda = 2$, and so the vertical lines C_2 and C_5 intersect the real axis at -1 and 1, respectively. The only change in the proof occurs in the calculation of the integral along the arc $C_{\tau_\lambda} = C_{\tau_2}$ and then letting ϵ tend to 0. A simple calculation shows that

$$\int_{C_{\tau_2}} \frac{f'(\tau)}{f(\tau)}\,d\tau = \int_{C_{\tau_2}} \frac{\frac{d}{d\tau}\{f(\tau)(\tau+1)^k\}}{f(\tau)(\tau+1)^k}\,d\tau - \int_{C_{\tau_2}} \frac{\frac{d}{d\tau}\{(\tau+1)^k\}}{(\tau+1)^k}\,d\tau.$$

Now,

$$\int_{C_{\tau_2}} \frac{\frac{d}{d\tau}\{(\tau+1)^k\}}{(\tau+1)^k}\,d\tau = ik\Delta_{C_{\tau_2}}(\arg),$$

which clearly tends to 0 as ϵ tends to 0, since the line $x = -1$ and the circle $|\tau| = 1$ are tangent at $\tau = -1$.

We now make the change of variable $t = \exp(2\pi i z)$, where $z = z(\tau)$ is given by (6.1). Then,

$$f(\tau)(\tau+1)^k = \sum_{n+\rho \geq 0} c_n e^{2\pi i(n+\rho)z} = t^{n_{-1}} \sum_{n=0}^{\infty} c_{m+n} t^n = t^{n_{-1}} g(t),$$

where m is defined as in Definition 6.3. Next, we examine what happens to C_{τ_2} under the transformation t. We have

$$t = e^{-2\pi i/(\tau+1)} = e^{-2\pi i(x+1-iy)/\{(x+1)^2+y^2\}}.$$

If $\tau = -1 + iy$,

$$t = e^{-2\pi/y}.$$

As $\tau \in C_{\tau_2}$ tends from $-1+iy$ to the unit circle, the argument of t decreases. When τ reaches the unit circle, then $x^2 + y^2 = 1$ and

$$t = e^{-2\pi i(x+1-iy)/(2x+2)} = -e^{-\pi y/(x+1)}.$$

Thus, as τ traverses C_{τ_2}, t traverses a simple path Γ in the lower half-plane beginning on the positive real axis and terminating on the negative real axis. Thus,

$$\int_{C_{\tau_2}} \frac{\frac{d}{d\tau}\{f(\tau)(\tau+1)^k\}}{f(\tau)(\tau+1)^k}\,d\tau = \int_\Gamma \frac{\frac{d}{dt}(t^{n_{-1}}g(t))}{t^{n_{-1}}g(t)}\,dt$$
$$= \int_\Gamma \left\{\frac{n_{-1}}{t} + \frac{g'(t)}{g(t)}\right\} dt.$$

Since $g(t)$ is analytic and $g(0) \neq 0$, we deduce that, by the argument principle,

$$\lim_{\epsilon\to 0} \frac{1}{2\pi i}\int_\Gamma \left\{\frac{n_{-1}}{t} + \frac{g'(t)}{g(t)}\right\} dt = \frac{n_{-1}}{2\pi}\Delta_\Gamma(\arg)$$
$$= \frac{n_{-1}}{2\pi}(-\pi) = -\frac{1}{2}n_{-1}.$$

We obtain the same result for the arc C_{τ_2+2}, so combining the calculations as in the proof of Lemma 5.1, we arrive at (6.11).

The second part of the lemma follows as in the proof of Corollary 5.2.

□

Remark 6.2. An analogue of Corollary 5.3(b) for $\lambda = 2$ follows directly from Lemma 6.1: If $k < 0$, then

$$M_0(2,k,\gamma) = M_1(2,k,\gamma) = \{0\}.$$

This can also be proved by a modification of the argument given in Chapter 5 to derive Corollary 5.3(b).

Corollary 6.1. *Let*

$$m = 1 + [k/4]$$

and

$$m' = 1 + [(k-2)/4].$$

Then,

$$\dim M_1(2,k,1) \leq m \tag{6.13}$$

and

$$\dim M_1(2,k,-1) \leq m'. \tag{6.14}$$

Proof. From Lemma 6.1, $n_\infty < m$. Recalling that, since $\lambda = 2$, the angle $\theta = 0$ in this case, we proceed exactly as in the proof of Corollary 5.1 to arrive at (6.13).

If $\gamma = -1$, then from Lemma 6.1, $n_i \geq 1$. Hence,

$$n_\infty \leq \frac{k}{4} - \frac{1}{2} < m'.$$

Proceed as in the proof of Corollary 5.1 to arrive at (6.14). □

Example 6.1. From Corollary 6.1,

$$\dim M_1\left(2, \frac{1}{2}, 1\right) \leq 1.$$

Since $M_0\left(2, \frac{1}{2}, 1\right) \subseteq M_1\left(2, \frac{1}{2}, 1\right)$, there is at most one Dirichlet series of signature $\left(2, \frac{1}{2}, 1\right)$. But from (1.2), $\zeta(2s)$ has signature $\left(2, \frac{1}{2}, 1\right)$. Thus, we have proved Hecke's version of Hamburger's result to which we referred in the Introduction, namely, that there is exactly one linearly independent solution to the functional equation satisfied by $\zeta(s)$.

We note here that the uniqueness of the solution to this functional equation is radically disturbed by (apparently innocent) modifications to the auxiliary conditions [62, Theorem 1]. See the application of Theorem 7.1 following Remark 7.4.

Example 6.2. It is known that the Dedekind zeta function for $K = \mathbb{Q}(i)$,

$$\zeta_K(s) = \frac{1}{4}\zeta(s, Q),$$

where $Q(m,n) = m^2 + n^2$, has signature $(2, 1, 1)$. (See the example in Chapter 2, following Definition 2.1.) But by Corollary 6.1, there exists at most one Dirichlet series of signature $(2, 1, 1)$. Thus $\zeta_K(s)$ is the unique solution to its functional equation.

We could now proceed as in Chapter 5. From the Riemann mapping theorem there exists a function $g(\tau)$ which maps $B(2)$ conformally onto $\mathcal{H}$ such that $g(-1) = 0$, $g(i) = 1$, and $g(i\infty) = \infty$. We would then construct functions f_{-1}, f_i, and f_∞ from which we could find a basis for $M_1(2, k, \gamma)$. This is, in fact, the method used by Hecke [48, pp. 24–28]. However, we follow a slightly shorter, more constructive method given by A. Ogg [87].

Lemma 6.2. [105, p. 406] *Let f be of bounded variation on $[0, 1]$. Then,*

$$\frac{1}{2}\{f(x+0) + f(x-0)\} = \frac{1}{2}a_0 + \sum(a_n \cos(2\pi nx) + b_n \sin(2\pi nx)),$$

where

$$\frac{1}{2}a_n = \int_0^1 f(x)\cos(2\pi n x)dx \quad \textit{and} \quad \frac{1}{2}b_n = \int_0^1 f(x)\sin(2\pi n x)dx.$$

Lemma 6.3. (Poisson summation formula) *Let f be of bounded variation on $[A, B]$, where A, $B \in Z$. Then,*

$$\frac{1}{2}\sum_{n=A}^{B}{}'\{f(n+0)+f(n-0)\} = \lim_{N\to\infty}\sum_{\nu=-N}^{N}\int_A^B f(x)e^{2\pi i\nu x}\,dx\,,$$

where the $'$ indicates that for $n = A$ only the term $\frac{1}{2}f(A+0)$ is counted and for $n = B$ only the term $\frac{1}{2}f(B-0)$ is counted.

The form and proof of Poisson's summation formula that we give here can be found in [27, p. 16]. See [60, pp. 39–40] for another proof.

In the sequel we write $\sum\limits_{\nu=-\infty}^{\infty}$ for $\lim\limits_{\nu\to\infty}\sum\limits_{\nu=-N}^{N}$.

Proof of Lemma 6.3. Let $f_1(x) = f(x)$, $0 \le x < 1$, and extend the definition of f_1 to $(-\infty, \infty)$ by periodicity. Apply Lemma 6.2 to f_1 and choose $x = 0$. Then,

$$\begin{aligned}\frac{1}{2}\{f(0+0)+f(1-0)\} &= \frac{1}{2}\{f_1(0+0)+f_1(0-0)\}\\ &= \frac{1}{2}\,a_0 + \sum a_\nu\\ &= \sum_{\nu=-\infty}^{\infty}\int_0^1 f(x)\cos(2\pi\nu x)dx.\end{aligned}$$

Replacing $f(x)$ by $f(x+n)$, where $n \in \mathbb{Z}$, we find that

$$\begin{aligned}\frac{1}{2}\{f(n+0)+f(n+1-0)\} &= \sum_{\nu=-\infty}^{\infty}\int_0^1 f(x+n)\cos(2\pi\nu x)dx\\ &= \sum_{\nu=-\infty}^{\infty}\int_n^{n+1} f(x)\cos(2\pi\nu x)dx.\end{aligned}$$

Now sum on n, $A \le n \le B-1$, to deduce that

$$\frac{1}{2}\sum_{n=A}^{B}{}' \{f(n+0)+f(n-0)\} = \sum_{\nu=-\infty}^{\infty}\int_A^B f(x)\cos(2\pi\nu x)dx$$
$$= \sum_{\nu=-\infty}^{\infty}\int_A^B f(x)e^{2\pi i\nu x}dx,$$

as the terms involving $\sin(2\pi\nu x)$ and $\sin(-2\pi\nu x)$ cancel. □

Theorem 6.2. *The classical theta function* $\theta(\tau) \in M_0\left(2, \frac{1}{2}, 1\right)$. *Furthermore, the only zero of* $\theta(\tau)$ *on* $\overline{B(2)}$ *(up to equivalence under* $G(2)$*) is at* $\tau = -1$*, and the order of the zero is* $1/8$.

Proof. Recall that $\theta(\tau)$ is defined at the beginning of Chapter 1. Since (i) and (iii) of Definition 2.2 are clearly satisfied, to prove the first part of the theorem we need only establish (1.1). It is sufficient to prove (1.1) for $\tau = iy$, $y > 0$, for then the general result will follow by analytic continuation.

By Lemma 6.3,

$$\sum_{n=-N}^{N}{}' e^{-\pi n^2/y} = \sum_{\nu=-\infty}^{\infty}\int_{-N}^{N} e^{-\pi x^2/y+2\pi i\nu x}\,dx \tag{6.15}$$
$$= \sum_{\nu=-\infty}^{\infty}\left\{\int_{-\infty}^{\infty} - \int_{|x|\ge N}\right\} e^{-\pi x^2/y+2\pi i\nu x}\,dx.$$

We now want to show that as N tends to ∞,

$$\sum_{\nu=-\infty}^{\infty}\int_{|x|\ge N} e^{-\pi x^2/y+2\pi i\nu x}\,dx$$

tends to 0. Upon integrating by parts twice, we find that

$$\int_N^\infty e^{-\pi x^2/y}\cos(2\pi\nu x)dx = -\frac{1}{2\pi\nu}\int_N^\infty \frac{d}{dx}\{e^{-\pi x^2/y}\}\sin(2\pi\nu x)dx$$
$$= \frac{N}{2\pi y\nu^2}e^{-\pi N^2/y} - \frac{1}{4\pi^2\nu^2}\int_N^\infty \frac{d^2}{dx^2}\{e^{-\pi x^2/y}\}\cos(2\pi\nu x)dx.$$

Hence,

$$\begin{aligned}
&\left| \sum_{\substack{\nu=-\infty \\ \nu \neq 0}}^{\infty} \int_{|x| \geq N} e^{-\pi x^2/y + 2\pi i \nu x} dx \right| \\
&\leq \frac{N}{\pi y} e^{-\pi N^2/y} \sum \nu^{-2} + \frac{1}{2\pi^2} \int_N^{\infty} \frac{d^2}{dx^2} \{e^{-\pi x^2/y}\} dx \sum \nu^{-2} \\
&= o(1),
\end{aligned}$$

as N tends to ∞. Thus, from our calculations above and (6.15), upon letting N tend to ∞, we deduce that

$$\begin{aligned}
\sum_{n=-\infty}^{\infty} e^{-\pi n^2/y} &= \sum_{\nu=-\infty}^{\infty} \int_{-\infty}^{\infty} e^{-\pi x^2/y + 2\pi i \nu x} dx \qquad (6.16) \\
&= y \sum_{\nu=-\infty}^{\infty} \int_{-\infty}^{\infty} e^{-\pi u^2 y + 2\pi i \nu y u} du,
\end{aligned}$$

where we have replaced x by uy. Apply Cauchy's theorem by integrating $\exp(-\pi z^2 y)$ with respect to z around the rectangle with vertices $\pm R$ and $\pm R - i\nu$ and then letting R tend to ∞. We immediately find that

$$\int_{-\infty}^{\infty} e^{-\pi(u - i\nu)^2 y} du = \int_{-\infty}^{\infty} e^{-\pi u^2 y} du. \qquad (6.17)$$

Now,

$$\int_{-\infty}^{\infty} e^{-\pi u^2 y} du = y^{-\frac{1}{2}} \int_{-\infty}^{\infty} e^{-\pi v^2} dv = y^{-\frac{1}{2}} B, \qquad (6.18)$$

say. Since $-\pi u^2 y + 2\pi i \nu y u = -\pi(u - i\nu)^2 y - \pi \nu^2 y$, we find that (6.16)–(6.18) yield

$$\sum_{n=-\infty}^{\infty} e^{-\pi n^2/y} = B y^{\frac{1}{2}} \sum_{\nu=-\infty}^{\infty} e^{-\pi \nu^2 y},$$

or

$$\theta(i/y) = B y^{\frac{1}{2}} \theta(iy).$$

Put $y = 1$. Since, clearly, $\theta(i) \neq 0$, we find that $B = 1$. This concludes the proof of (1.1) and the first part of the theorem.

To prove the second part of the theorem, we have from (6.11), since $M_0\left(2, \frac{1}{2}, 1\right) \subseteq M_1\left(2, \frac{1}{2}, 1\right)$,

$$N + n_\infty + \frac{1}{2} n_i + n_{-1} = \frac{1}{8}.$$

Since N, n_∞, and n_i are integers, the only possible way this can be satisfied is if $N = n_\infty = n_i = 0$ and $n_{-1} = 1/8$, and the proof is complete □

Remark 6.3. The important functional equation (1.2) for $\zeta(s)$, which has motivated much of the material in this volume, (finally) follows directly from Theorem 6.2 and the Main Correspondence Theorem, Theorem 2.1. For the derivation, see Example 6.4 following the proof of Theorem 6.3.

Theorem 6.3. *We have* $M_0(2, k, \gamma) = M_1(2, k, \gamma)$. *Furthermore,*

$$\dim M_0(2, k, \gamma) = 1 + [(k + \gamma - 1)/4].$$

Note that in contrast to the case $\lambda < 2$ (see Theorem 5.6), there are no restrictions on $k > 0$ when $\lambda = 2$. The underlying reason for this difference is that $T_1 T_3 = \left(\begin{smallmatrix} 0 & -1 \\ 1 & \lambda \end{smallmatrix}\right)$ is elliptic for $\lambda < 2$ and parabolic for $\lambda = 2$. (See Chapter 3.) Thus, with $\lambda < 2$, the fixed point τ_λ of $T_1 T_3$ is in $\mathcal{H}$, while $\tau_2 = -1$, a rational point on the real line. Consequently, $T_1 T_3$ has finite order when $\lambda < 2$, but not when $\lambda = 2$. Indeed, the expression for $(T_1 T_3)^n$ given at the beginning of the proof of Theorem 5.2 shows that with $p = 1$,

$$(T_1 T_3)^q = -I.$$

This, together with $T^2 = -I$, gives two independent relations in $G(\lambda)$, $\lambda = 2\cos \pi/q$, while $G(2)$ has only the single relation $T^2 = -I$. The extra relation in the former cases imposes the additional restrictions on k by forcing conditions upon the multiplier system of an automorphic form with respect to $G(\lambda)$, when $\lambda < 2$.

Proof of Theorem 6.3. We show that

$$\dim M_0(2, k, 1) \geq 1 + [k/4].$$

Since $M_0\left(2, \frac{1}{2}, 1\right) \subseteq M_1\left(2, \frac{1}{2}, 1\right)$, it follows immediately from Corollary 6.1 that $M_0(2, k, 1) = M_1(2, k, 1)$ and that

$$\dim M_0(2, k, 1) = 1 + [k/4].$$

A similar argument will establish the result in the case $\gamma = -1$.

The proof of the aforementioned lower bound on $\dim M_0(2,k,1)$ entails consideration of arbitrary positive real powers of $\theta(\tau)$. Since $\theta(\tau) \neq 0$ and analytic on $\mathcal{H}$, and since $\mathcal{H}$ is simply connected, we can define an analytic function $\log \theta(\tau)$ on $\mathcal{H}$. Since $\theta(\tau)$ can be represented as a power series in $z = \exp(\pi i \tau)$ with nonzero constant term, $\log \theta(\tau)$ can be represented as a power series in z. For any $k > 0$, define

$$\theta^{2k}(\tau) = e^{2k \log \theta(\tau)},$$

where the logarithm is, as usual, defined by the assumption $\log x \in \mathbb{R}$ for $x > 0$. In this case, the assumption amounts to $\log \theta(iy) \in \mathbb{R}$ for $y > 0$. This definition gives rise naturally to a power series representation of $\theta^{2k}(\tau)$ in the variable $z = \exp(\pi i \tau)$. Thus, (i) of Definition 2.2 is satisfied. Because of Lemma 2.2 and the fact that $\theta(\tau)$ satisfies condition (iii) of Definition 2.2, it follows that $\theta^{2k}(\tau)$ obeys condition (iii) as well. We also have

$$\theta^{2k}(-1/\tau) = \epsilon(\tau/i)^k \theta^{2k}(\tau)$$

for some complex number ϵ, where $|\epsilon| = 1$. Letting $\tau = i$ and using the fact that $\theta(i) \neq 0$, we conclude immediately that $\epsilon = 1$. Hence $\theta^{2k}(\tau) \in M_0(2,k,1)$, and, by Theorem 6.2, the only zero (up to $G(2)$-equivalence) of $\theta^{2k}(\tau)$ on $\overline{B(2)}$ is at $\tau = -1$ with order $k/4$.

Recall that in Theorem 5.9 we showed that the Eisenstein series $E_4(\tau) \in M_0(1,4,1)$. Since $M_0(1,4,1) \subseteq M_0(2,4,1) \subseteq M_1(2,4,1)$, $E_4(\tau)$ is in $M_1(2,4,1)$ as well. In the lines immediately preceding Definition 5.4, we observed further that the only zero (up to $G(1)$-equivalence) of $E_4(\tau)$ on $\overline{B(1)}$ is a simple one at $\tau_1 = \exp(2\pi i/3)$. Now, from Lemma 6.1,

$$N + n_\infty + \frac{1}{2} n_i + n_{-1} = 1.$$

It follows that τ_1 is the only zero (up to $G(2)$-equivalence) of $E_4(\tau)$ on $\overline{B(2)}$ as well. To see this, we observe the potentially elusive distinction between the order of the zero of $E_4(\tau)$ at τ_1 with respect to $G(1)$ and the order of the same zero with respect to $G(2)$. The former is measured in the local variable $\{(\tau - \tau_1)/(\tau - \bar{\tau}_1)\}^3$ with respect to $G(1)$, since τ_1 is a fixed point of $S^{-1}T$, an element in $G(1)$ of order 3. The latter is measured in the standard planar variable, since τ_1 is not a fixed point of $G(2)$. Thus the latter order is three times the former.

Consider the functions $E_4^\nu \theta^{2k-8\nu}$, with $\nu \in Z$, $\nu \geq 0$, and $2k - 8\nu \geq 0$. By the properties of θ^{2k} and E_4 that we have already established, these functions lie in $M_0(2,k,1)$. By the inequality imposed upon ν, $0 \leq \nu \leq k/4$, so there are precisely $1+[k/4]$ such functions. On the other hand, $E_4^\nu \theta^{2k-8\nu}$ has a zero of order exactly ν at $\tau_1 = \exp(2\pi i/3)$ with respect to $G(2)$, so that the elements of the set $\{E_4^\nu \theta^{2k-8\nu} : 0 \leq \nu \leq [k/4]\}$ are linearly independent. Thus, $\dim M_0(2,k,1) \geq 1 + [k/4]$, and the proof of Theorem 6.3 is complete in the case $\gamma = 1$.

Let $\gamma = -1$. We have already observed that the only zero of E_4 on $\overline{B(2)}$ is at τ_1. Thus, since $E_4(i) \neq 0$, we can choose α such that

$$g := \theta^8 - \alpha E_4$$

has a zero at $\tau = i$. As $\theta(i) \neq 0$, $\alpha \neq 0$. Since $E_4(-1) \neq 0$ and $\theta(-1) = 0$, $g(-1) \neq 0$. Now $\theta \in M_0\left(2, \frac{1}{2}, 1\right)$ and $E_4 \in M_0(2,4,1)$, so by Lemma 2.2, $g \in M_0(2,4,1)$. It follows that we may apply Lemma 6.1 to g, with the result

$$N + n_\infty + \frac{1}{2} n_i = \frac{4}{4} = 1.$$

Since $g(i) = 0$ by construction, $n_i > 0$. This last equation then yields the values $n_i = 2$, $N = n_\infty = 0$. That is, in the standard planar variable $\tau - i$, g has a double zero at $\tau = i$ and no other zeros in $\overline{B(2)}$.

Since g has a double zero at $\tau = i$, we may define an analytic square root $h = g^{\frac{1}{2}}$ in a neighborhood of $\tau = i$ and all its equivalent points on $\mathcal{H}$. Since g has no other zeros on $\mathcal{H}$, we may define an analytic square root in a neighborhood of any other point in $\mathcal{H}$ as well. As $\mathcal{H}$ is simply connected, by the monodromy theorem, h is analytic on $\mathcal{H}$.

We claim that $h \in M_0(2,2,-1)$. We apply Lemma 2.2 to find that h satisfies condition (iii) of Definition 2.2. To check that (i) holds, recall that θ and E_4 both satisfy (i), so that, in particular, $g(\tau+2) = g(\tau)$. From this it follows directly that $h(\tau+2) = \pm h(\tau)$. We show that $h(\tau+2) = h(\tau)$. We have observed above that $g(i\infty) \neq 0$ (i.e., $n_\infty = 0$) so that $h(i\infty) \neq 0$. Furthermore $h(i\infty) = \lim_{y\to\infty} h(iy)$ is finite. Thus, if $h(\tau+2) = -h(\tau)$, then a simple argument, which we have invoked earlier, shows that

$$h(\tau) = \sum_{n=0}^{\infty} a_n e^{\pi i (n+\frac{1}{2})\tau},$$

for $\tau \in \mathcal{H}$. But this implies that $h(i\infty) = 0$, a contradiction. In fact, then,

$$h(\tau) = \sum_{n=0}^{\infty} a_n e^{\pi i n \tau}, \quad \tau \in \mathcal{H},$$

so h satisfies condition (i) of Definition 2.2 for membership in $M_0(2, 2, -1)$.

It remains to verify condition (ii). Now, since $g \in M_0(2, 4, 1)$, we have $g(-1/\tau) = (\tau/i)^4 g(\tau)$, so that

$$h(-1/\tau) = \pm(\tau/i)^2 h(\tau). \tag{6.19}$$

Since $h(\tau)$ has a simple zero at $\tau = i$, we may write

$$h(\tau) = (\tau - i) h_1(\tau),$$

where $h_1(i) \neq 0$. Then

$$h(-1/\tau) = -\frac{i}{\tau}(\tau - i) h_1(-1/\tau).$$

Thus, (6.19) may be written as

$$-\frac{i}{\tau} h_1(-1/\tau) = \epsilon(\tau/i)^2 h_1(\tau).$$

Letting $\tau = i$, we find that $\epsilon = -1$. Hence, $h \in M_0(2, 2, -1)$.

Suppose now that $f_1 \in M_0(2, k-2, 1)$. Then, using Lemma 2.2 once more to verify condition (iii), we find that $f = f_1 h \in M_0(2, k, -1)$. Clearly, then, the map $f_1 \to f = f_1 h$ is one-to-one from $M_0(2, k-2, 1)$ into $M_0(2, k, -1)$, so that

$$\dim M_0(2, k, -1) \geq \dim M_0(2, k-2, 1) = 1 + [(k-2)/4],$$

by the first part of the theorem. But by Corollary 6.1, it follows that

$$\dim M_0(2, k, -1) \leq \dim M_1(2, k, -1) \leq 1 + [(k-2)/4],$$

so that $\dim M_0(2, k, -1) = 1 + [(k-2)/4] = \dim M_1(2, k, -1)$. This completes the proof of Theorem 6.3. □

From the definition of h in the proof of Theorem 6.3, it follows that $h^2 = \theta^8 - \alpha E_4$, with $\alpha \neq 0$. Thus E_4 is a polynomial in θ and h. The proof of Theorem 6.3 then shows that any element of $M_0(2, k, \gamma)$ is a polynomial in θ and h. This has great significance for Dirichlet series of signature $(2, k, \gamma)$, for each such series can be obtained from those corresponding to the signatures $\left(2, \frac{1}{2}, 1\right)$ and $(2, 2, -1)$.

We conclude this chapter with two number-theoretic examples.

Example 6.3. If k is a positive integer, then for $\tau \in \mathcal{H}$,

$$\theta^k(\tau) = \sum_{n_1,\ldots,n_k=-\infty}^{\infty} e^{\pi i(n_1^2+\cdots+n_k^2)\tau} = 1 + \sum r_k(n) e^{\pi i n\tau},$$

where $r_k(n)$ denotes the number of representations of n as the sum of k squares. If $k < 8$, there is only one Dirichlet series, $\sum r_k(n)n^{-s}$, with signature $(2, \frac{k}{2}, 1)$, as follows rather directly from the proof of Theorem 3 of Chapter 5 of [60]. (See page 81 of [60].) Of course, this is likewise an immediate consequence of Theorem 6.3. This is a very important fact, for if we can find another series $\sum a_k(n)n^{-s}$ with the same signature, then $r_k(n)$ is a multiple of $a_k(n)$. Thus, one might obtain formulas for $r_k(n)$. For an early historical account of formulas that have been found for $r_k(n)$, see [4], and for recent extremely comprehensive surveys, see the two papers [78] and [79] by S. C. Milne.

Example 6.4. We begin by deriving the functional equation (1.2) from Theorem 6.2. The latter shows that condition (i) of Theorem 2.1 is satisfied with $\lambda = 2$, $k = \frac{1}{2}$, $\gamma = 1$, and $f(\tau) = g(\tau) = \theta(\tau)$. Furthermore, in the notation of Theorem 2.1, $a_0 = b_0 = 1$ and $a_n = b_n = 2$, if $n = m^2$, $m \in \mathbb{Z}^+$; $a_n = b_n = 0$, otherwise. Thus

$$\varphi(s) = \psi(s) = 2\sum_{m=1}^{\infty} \frac{1}{(m^2)^s} = 2\zeta(2s)$$

satisfies (2.1). That is to say,

$$\pi^{-s}\Gamma(s)\zeta(2s) = \pi^{s-1/2}\Gamma\left(\tfrac{1}{2} - s\right)\zeta(1-2s);$$

replacing s by $s/2$, we obtain

$$\pi^{-s/2}\Gamma(s/2)\zeta(s) = \pi^{(s-1)/2}\Gamma(\{1-s\}/2)\zeta(1-s),$$

the functional equation (1.2).

Let k be an even positive integer. From the foregoing functional equation of $\zeta(s)$,

$$\begin{aligned}\pi^{-s+(k-1)/2}&\Gamma(s/2)\Gamma(\{s-k+1\}/2)\zeta(s)\zeta(s-k+1) \\ &= \pi^{s-(k+1)/2}\Gamma(\{1-s\}/2)\Gamma(\{k-s\}/2)\zeta(1-s)\zeta(k-s).\end{aligned} \tag{6.20}$$

From Legendre's duplication formula,

$$\Gamma(s) = \pi^{-1/2} 2^{s-1} \Gamma(s/2) \Gamma(\{1+s\}/2),$$

we find that

$$\Gamma(s/2)\Gamma(\{s-k+1\}/2) = \frac{\pi 2^{1-2s+k}\Gamma(s)\Gamma(s-k+1)}{\Gamma(\{s+1\}/2)\Gamma(\{s-k+2\}/2)}. \tag{6.21}$$

After three applications of the reflection formula,

$$\Gamma(s)\Gamma(1-s) = \frac{\pi}{\sin(\pi s)},$$

(6.21) becomes

$$\begin{aligned}
&\Gamma(s/2)\Gamma(\{s-k+1\}/2) \\
&= \frac{2^{1-2s+k}\Gamma(s)\Gamma(\{1-s\}/2)\Gamma(\{k-s\}/2)\sin\{\pi(s+1)/2\}\sin\{\pi(s-k+2)/2\}}{\Gamma(k-s)\sin\{\pi(k-s)\}} \\
&= \frac{2^{-2s+k}(-1)^{k/2}\Gamma(s)\Gamma(\{1-s\}/2)\Gamma(\{k-s\}/2)}{\Gamma(k-s)}.
\end{aligned} \tag{6.22}$$

Substituting (6.22) into (6.20), we find that

$$(2\pi)^{-s}\Gamma(s)\zeta(s)\zeta(s-k+1) = (-1)^{k/2}(2\pi)^{s-k}\Gamma(k-s)\zeta(1-s)\zeta(k-s).$$

Thus, we have shown that $\zeta(s)\zeta(s-k+1)$ has signature $(1, k, (-1)^{k/2})$, provided that we show that $\zeta(s)\zeta(s-k+1)$ is analytic except for a simple pole at $s = k$. Since $\zeta(s)$ has a simple pole at $s = 1$, it is clear that $\zeta(s)\zeta(s-k+1)$ has a simple pole at $s = k$. Now, from the functional equation for $\zeta(s)$ it is easy to see that $\zeta(-n) = 0$ for every even positive integer n. Thus, if $k > 2$, $\zeta(2-k) = 0$. Hence $\zeta(s)\zeta(s-k+1)$ is analytic at $s = 1$ if $k > 2$. Thus, we have proved Corollary 5.5 once again.

For several examples of Dirichlet series of signature $(2, k, \gamma)$ when $k > 0$ is integral, see [48, pp. 29–31].

Chapter 7

Bochner's generalization of the main correspondence theorem of Hecke, and related results

Bochner's generalization [17], [20, pp. 665–696] has previously been mentioned in passing in Chapter 2, following the proof of Theorem 2.1, the Hecke correspondence theorem. We take up the generalization in some detail here, both for its intrinsic interest and because it has proved very useful in applications to the theory of Dirichlet series. (See [7], [61], [62], [66], and the remarks following the proof of Theorem 7.1 below.) After he published [17], [20, pp. 665–696], in the next few years Bochner returned to his ideas from [17], reformulated his main correspondence theorem, and derived further consequences. In particular, see his papers [18], [21], [19], [20, pp. 697–700, 715–739, 763–775]; the paper [21] was coauthored with K. Chandrasekharan.

In formulating the generalized correspondence theorem it is useful to introduce a special class of functions, the "log-polynomial sums."

Definition 7.1. A *log-polynomial sum* (LPS) $q(\tau)$ is a finite sum of the form

$$q(\tau) = \sum_{1 \le j \le L} (\tau/i)^{\alpha_j} \sum_{0 \le t \le M(j)} \beta(j,t) \log^t(\tau/i), \tag{7.1}$$

with L, $M(j) \in \mathbb{Z}$ and α_j, $\beta(j,t)$ complex constants.

Note that $q(\tau)$ is holomorphic in the slit plane $\mathbb{C} - \{it : t \le 0\}$. In particular, $q(\tau)$ is holomorphic in $\mathcal{H}$. It is worth noting that LPS's are "residual functions" in the sense of Bochner [17], [20, pp. 665–696].

Bochner's generalization follows.

Theorem 7.1. *Let λ_1, $\lambda_2 > 0$, $k \in \mathbb{R}$, and $\gamma \in \mathbb{C}$. Let*

$$f(\tau) = \sum_{n=0}^{\infty} a_n e^{2\pi i n\tau/\lambda_1}, \qquad g(\tau) = \sum_{n=0}^{\infty} b_n e^{2\pi i n\tau/\lambda_2} \tag{7.2}$$

be nonconstant exponential series, such that the sequences $\{a_n\}$, $\{b_n\}$ *satisfy the growth condition*

$$a_n, b_n = O(n^\rho), \qquad n \to \infty, \qquad \rho > 0.$$

As in Theorem 2.1, *put*

$$\Phi(s) = \left(\frac{2\pi}{\lambda_1}\right)^{-s} \Gamma(s) \sum a_n n^{-s} = \int_0^\infty \{f(iu) - a_0\} u^{s-1} du, \tag{7.3}$$

$$\Psi(s) = \left(\frac{2\pi}{\lambda_2}\right)^{-s} \Gamma(s) \sum b_n n^{-s} = \int_0^\infty \{g(iu) - b_0\} u^{s-1} du.$$

Then the following two conditions are equivalent.

(A) $f(\tau)$ *and* $g(\tau)$ *satisfy the (generalized) modular transformation equation*

$$(\tau/i)^{-k} f(-1/\tau) = \gamma g(\tau) + q(\tau), \tag{7.4}$$

where $q(\tau)$ *is an LPS.*

(B) $\Phi(s)$ *and* $\Psi(s)$ *have meromorphic continuations to the entire s-plane, each with at most a finite number of poles in* $\mathbb{C}$. *Furthermore,* $\Phi(s)$ *and* $\Psi(s)$ *satisfy the functional equation*

$$\Phi(k - s) = \gamma \Psi(s). \tag{7.5}$$

Finally, there exists $T_0 > 0$ *such that* $\Phi(s)$ *remains bounded in each "lacunary" vertical strip (LVS) of the form* $\sigma_1 \le \sigma \le \sigma_2$, $|t| \ge T_0$. *(Here,* $s = \sigma + it$.*)*

Remark 7.1. 1. To compare Theorem 7.1 with Theorem 2.1, note that we have here interchanged the roles of f and g, and consequently the roles of Φ and Ψ, as well. In this generalization λ_1, λ_2 may be distinct positive numbers whereas in Theorem 2.1 $\lambda = \lambda_1 = \lambda_2$, even when $f \neq g$.

2. The most important feature of the generalization is the presence of the LPS $q(\tau)$ in (7.4). (Recall that $q = 0$ in Theorem 2.1). It is striking that the presence of q in (7.4) makes no change in the functional equation connecting Φ and Ψ. The influence of q makes itself felt rather in the polar singularities of Φ and Ψ. The details will emerge in the proof that (A) implies (B). The structure of the proof is the same as that of Theorem 2.1.

In the proof of Theorem 7.1 we require two preliminary results.

Lemma 7.1. *Let $\alpha \in \mathbb{C}$, with $\operatorname{Re}\alpha < -1$, and let $j \in \mathbb{Z}$, $j \geq 0$. Then,*

$$\int_1^\infty u^\alpha (\log u)^j du = (-1)^{j+1} j!(\alpha+1)^{-j-1}. \tag{7.6}$$

Proof. Integrate by parts j times or apply induction. □

Proposition 7.1. *Assume $f(\tau)$ is nonconstant and has the form*

$$f(\tau) = \sum_{n=0}^\infty a_n e^{2\pi i n\tau/\lambda}, \qquad \tau \in \mathcal{H}, \qquad \lambda > 0.$$

Assume as well that f satisfies the transformation law (7.4), with $g = f$ and $q(\tau)$ of the form (7.1). Then $\gamma^2 = 1$ and

$$q(-1/\tau) = -e^{-\pi i k/2}\gamma\tau^k q(\tau), \qquad \tau \in \mathcal{H}. \tag{7.7}$$

Proof. Applying (7.4) twice yields

$$f(\tau) = \gamma^2 f(\tau) + \gamma q(\tau) + (i/\tau)^k q(-1/\tau),$$

so that

$$f(\tau)(1-\gamma^2) = \gamma q(\tau) + (i/\tau)^k q(-1/\tau).$$

But the left-hand side is periodic with period λ, while (as is easily checked) the right-hand side is a LPS and thus not periodic unless it is constant. Since $f(\tau)$ is nonconstant, the left-hand side is constant only when $\gamma^2 = 1$, and (7.7) follows. □

In order to emphasize the role of the LPS in the generalization, we begin our treatment of Theorem 7.1 with a proof that (A) implies (B) in the special case $f = g$ (hence $\lambda_1 = \lambda_2$ and $\Phi = \Psi$), thus suppressing the less important generalization to distinct f and g.

Proof that (A) implies (B) (assuming $f = g$). Since $\lambda_1 = \lambda_2$, we set $\lambda = \lambda_1 = \lambda_2$. As in the proof of Theorem 2.1, we find that

$$\Phi(s) = \int_1^\infty \{f(iu) - a_0\}u^{s-1}du + \int_1^\infty \{f(i/u) - a_0\}u^{-s-1}du.$$

Apply (7.4) with $\tau = iu$, $u > 0$, and rearrange slightly to obtain

$$\int_1^\infty \{f(i/u) - a_0\}u^{-s-1}du = \gamma \int_1^\infty \{f(iu) - a_0\}u^{k-s-1}du$$
$$+ \gamma \int_1^\infty a_0 u^{k-s-1}du - \int_1^\infty a_0 u^{-s-1}du + \int_1^\infty q(iu)u^{k-s-1}du.$$

Thus, for $\sigma > \max(0, k)$,

$$\begin{aligned}\Phi(s) &= \int_1^\infty \{f(iu) - a_0\}u^{s-1}du + \gamma \int_1^\infty \{f(iu) - a_0\}u^{k-s-1}du \\ &\quad + a_0\left(\frac{\gamma}{s-k} - \frac{1}{s}\right) + \int_1^\infty q(iu)u^{k-s-1}du \\ &= E(s) + r(s) + \int_1^\infty q(iu)u^{k-s-1}du,\end{aligned}$$

with $E(s)$ equal to the sum of the first two integrals above and

$$r(s) = a_0\left(\frac{\gamma}{s-k} - \frac{1}{s}\right). \tag{7.8}$$

By the exponential decay of $f(iu) - a_0$ as $u \to +\infty$, it follows that $E(s)$ is entire. Also, by Proposition 7.1, $\gamma = \pm 1$, so that

$$E(k-s) = \gamma E(s), \qquad r(k-s) = \gamma r(s). \tag{7.9}$$

To continue the proof, we examine the remaining term

$$\mathcal{L}(s) := \int_1^\infty q(iu)u^{k-s-1}du. \tag{7.10}$$

By (7.7), we have the alternative representation

$$\mathcal{L}(s) = -\gamma \int_1^\infty q(i/u)u^{-s-1}du. \tag{7.11}$$

Note that both integrals (7.10) and (7.11) converge and are holomorphic for $\sigma = \operatorname{Re} s$ sufficiently large.

To apply Lemma 7.1 to (7.10), we single out a term in $q(iu)$ and suppress the indices j, t; by (7.1), such a term has the form $\beta u^\alpha(\log u)^t$, with α, $\beta \in \mathbb{C}$, $\beta \neq 0$, and $t \in \mathbb{Z}$, with $t \geq 0$. In (7.10) this gives rise to the term

$$\beta \int_1^\infty u^{\alpha+k-s-1}(\log u)^t du,$$

which by Lemma 7.1 equals $\beta(-1)^{t+1}t!(\alpha+k-s)^{-t-1}$, provided that $\sigma > \operatorname{Re}\alpha + k$. It follows that, for $\sigma > k + \max_{1\le j\le L}\{\operatorname{Re}\alpha_j\}$,

$$\mathcal{L}(s) = \sum_{1\le j\le L}\ \sum_{0\le t\le M(j)} \beta(j,t)(-1)^{t+1}t!(\alpha_j+k-s)^{-t-1}, \tag{7.12}$$

and the right-hand side of this equation provides the meromorphic continuation of $\mathcal{L}(s)$ to all of $\mathbb{C}$.

To derive the functional equation (7.5), it suffices, by the decomposition

$$\Phi(s) = E(s) + r(s) + \mathcal{L}(s)$$

and (7.9), to show that

$$\mathcal{L}(k-s) = \gamma\mathcal{L}(s), \qquad s\in\mathbb{C}. \tag{7.13}$$

From (7.12) it follows that

$$\mathcal{L}(k-s) = \sum_{1\le j\le L}\ \sum_{0\le t\le M(j)} \beta(j,t)(-1)^{t+1}t!(\alpha_j+s)^{-t-1}. \tag{7.14}$$

Next, recalculate $\mathcal{L}(s)$, this time employing the representation (7.11) in place of (7.10); the same method applied in deriving (7.12) shows that $\gamma\mathcal{L}(s)$ is the same as the right-hand side of (7.14), as long as $\sigma > \max_{1\le j\le L}\{-\operatorname{Re}\alpha_j\}$, and so (7.13) follows.

To complete the proof that (A) implies (B) we need only show that $\Phi(s)$ is bounded in a LVS. If we choose $T_0 = 1 + \max\{|\operatorname{Im}\alpha_j| : 1\le j\le L\}$ (say), then (7.8) and (7.12) imply that $r(s)$ and $\mathcal{L}(s)$ are bounded in the LVS, while the boundedness of $E(s)$ in the strip follows just as in the proof of Theorem 2.1, that is, by taking absolute values inside the two integrals whose sum is $E(s)$. This completes the proof that (A) implies (B). $\square$

Remark 7.2. The proof yields information about the poles of $\Phi(s)$: since $E(s) = \Phi(s) - r(s) - \mathcal{L}(s)$ is an entire function of s, (7.8) and (7.12) show that $\Phi(s)$ has poles at the points $\alpha_j + k$, and at 0 and k if $a_0 \ne 0$. On the other hand, the functional equation (7.13) implies that the poles of $\Phi(s)$ are located at the points $-\alpha_j$, and at 0 and k if $a_0\ne 0$. Thus, if $a_0 \ne 0$, we have the set equality

$$\{0, k, \alpha_j + k : 1\le j\le L\} = \{0, k, -\alpha_j : 1\le j\le L\}.$$

If $a_0 = 0$, the equality is

$$\{\alpha_j + k : 1 \leq j \leq L\} = \{-\alpha_j : 1 \leq j \leq L\}.$$

Rather than proceeding with a proof of the converse direction in Theorem 7.1 for the special case $f = g$, we continue by proving Theorem 7.1 in its full generality, in both directions.

Proof that (A) implies (B) (in general). Since we have removed the restriction $f = g$, Proposition 7.1 no longer applies. Thus the proof must be modified, but surprisingly little.

Assuming (A) and arguing as we did before, we obtain, for $\sigma = \operatorname{Re} s$ sufficiently large,

$$\Phi(s) = \int_1^\infty \{f(iu) - a_0\}u^{s-1}du + \gamma \int_1^\infty \{g(iu) - b_0\}u^{k-s-1} + \left(\frac{\gamma b_0}{s-k} - \frac{a_0}{s}\right) + \int_1^\infty q(iu)u^{k-s-1}du. \tag{7.15}$$

We require an analogous expression for $\Psi(s)$; in order to obtain this, we rewrite the transformation law (7.4) as

$$(\tau/i)^{-k}g(-1/\tau) = \gamma^{-1}f(\tau) - \gamma^{-1}(\tau/i)^{-k}q(-1/\tau). \tag{7.16}$$

Just as (7.15) follows from (7.4), (7.16) implies that

$$\Psi(s) = \int_1^\infty \{g(iu) - b_0\}u^{s-1}du + \gamma^{-1} \int_1^\infty \{f(iu) - a_0\}u^{k-s-1}du + \left(\frac{\gamma^{-1}a_0}{s-k} - \frac{b_0}{s}\right) - \gamma^{-1} \int_1^\infty q(i/u)u^{-s-1}du, \tag{7.17}$$

for σ sufficiently large.

With the possible exception of the integrals involving the period function $q(\tau)$, the right-hand sides of (7.15) and (7.17) are meromorphic in the entire s-plane. We apply Lemma 7.1 to carry out the meromorphic continuation into $\mathbb{C}$ of these two integrals as well. A single term $q(\tau)$ in the LPS has the form $\beta(\tau/i)^\alpha \log^t(\tau/i)$, and so by Lemma 7.1 the corresponding term in $\int_1^\infty q(iu)u^{k-s-1}du$ is

$$\beta \int_1^\infty u^\alpha (\log u)^t u^{k-s-1}du = \beta(-1)^{t+1}t!(\alpha + k - s)^{-t-1},$$

for σ sufficiently large. In the same way, the corresponding term in

$$-\gamma^{-1}\int_1^\infty q(i/u)u^{-s-1}\,du$$

is

$$\beta\gamma^{-1}(-1)^{t+1}t!(\alpha+s)^{-t-1}.$$

It follows that (7.15) and (7.17) can be written, respectively, as

$$\begin{aligned}\Phi(s) = &\int_1^\infty \{f(iu)-a_0\}u^{s-1}\,du + \gamma\int_1^\infty \{g(iu)-b_0\}u^{k-s-1}du\\ &+\left(\frac{\gamma b_0}{s-k}-\frac{a_0}{s}\right)+\mathcal{L}_1(s) \end{aligned}\tag{7.18}$$

and

$$\begin{aligned}\Psi(s) = &\int_1^\infty \{g(iu)-b_0\}u^{s-1}du + \gamma^{-1}\int_1^\infty \{f(iu)-a_0\}u^{k-s-1}du\\ &+\left(\frac{\gamma^{-1}a_0}{s-k}-\frac{b_0}{s}\right)+\mathcal{L}_2(s), \end{aligned}\tag{7.19}$$

where

$$\mathcal{L}_1(s) = \sum_{1\le j\le L}\ \sum_{0\le t\le M(j)} \beta(j,t)(-1)^{t+1}t!(\alpha_j+k-s)^{-t-1} \tag{7.20}$$

and

$$\mathcal{L}_2(s) = \gamma^{-1}\sum_{1\le j\le L}\ \sum_{0\le t\le M(j)} \beta(j,t)(-1)^{t+1}t!(\alpha+s)^{-t-1}. \tag{7.21}$$

From (7.18) and (7.20) we conclude that $\Phi(s)$ is meromorphic in $\mathbb{C}$; similarly, (7.19) and (7.21) imply that $\Psi(s)$ is meromorphic in $\mathbb{C}$. It follows further from the pairs of expressions (7.18), (7.20) and (7.19), (7.21) that $\Phi(s)$ and $\Psi(s)$ are bounded in each LVS of the form given in (B), for suitably large $T_0 > 0$.

Since both $\Phi(s)$ and $\Psi(s)$ are meromorphic in all of $\mathbb{C}$, the functional equation (7.5) is meaningful. To prove (7.5), we compare (7.18) and (7.19), obtaining

$$\Phi(k-s)-\gamma\Psi(s) = \mathcal{L}_1(k-s)-\gamma\mathcal{L}_2(s).$$

Thus, (7.5) will follow from

$$\mathcal{L}_1(k-s) = \gamma\mathcal{L}_2(s). \tag{7.22}$$

But, (7.22) is immediate from (7.20) and (7.21). This concludes the proof that (A) implies (B). □

Remark 7.3. By (7.18) and (7.20), the poles of $\Phi(s)$ are restricted to the set $\{0, k, \alpha_j + k; 1 \le j \le L\}$. In the same way, (7.19) and (7.21) imply that the poles of $\Psi(s)$ lie in the set $\{0, k, -\alpha_j : 1 \le j \le L\}$.

Proof that (B) implies (A). We turn to the proof of Theorem 7.1 in the converse direction, (B) implies (A). In (B) we assume that the functions $\Phi(s)$, $\Psi(s)$, defined by (7.3), have meromorphic continuations to the whole s-plane, each of them with at most a finite number of poles in $\mathbb{C}$. Further, assume that they are bounded in LVS's sufficiently removed from the real line and, finally, that they satisfy the functional equation (7.5).

As in the proof of Theorem 2.1, we begin by observing that for $y > 0$ and sufficiently large $d > 0$, we have both

$$g(iy) - b_0 = \frac{1}{2\pi i} \int_{(d)} \Psi(s) y^{-s} ds \tag{7.23}$$

and

$$f(iy) - a_0 = \frac{1}{2\pi i} \int_{(d)} \Phi(s) y^{-s} ds. \tag{7.24}$$

(Note that (7.3) defines Φ and Ψ as the Mellin transforms of $f - a_0$ and $g - b_0$, respectively, while the equations (7.24) and (7.23) express $f - a_0$ and $g - b_0$ as the inverse Mellin transforms of Φ and Ψ, respectively.) By assumption, there exist

$$P_1(s) = \sum_{1 \le j \le L} \sum_{1 \le t \le N(j)} \frac{b(j,t)}{(s - \delta_j)^t},$$

$$P_2(s) = \sum_{1 \le j \le L'} \sum_{1 \le t \le N'(j)} \frac{b'(j,t)}{(s - \eta_j)^t},$$

with δ_j, $\eta_j \in \mathbb{C}$, such that $\Phi(s) - P_1(s)$ and $\Psi(s) - P_2(s)$ are entire.

Applying the procedure in the proof of Theorem 2.1, we move the line of integration to $\sigma = -d$; (7.23) then implies that

$$g(iy) - b_0 = \frac{1}{2\pi i}\int_{(-d)} \Psi(s) y^{-s} ds + \sum_{1\le j\le L'} y^{-\eta_j} \sum_{1\le t\le N'(j)} b'(j,t)\frac{(-\log y)^{t-1}}{(t-1)!}, \tag{7.25}$$

whereas from (7.24) it follows that

$$f(iy) - a_0 = \frac{1}{2\pi i}\int_{(-d)} \Phi(s) y^{-s} ds + \sum_{1\le j\le L} y^{-\delta_j} \sum_{1\le t\le N(j)} b(j,t)\frac{(-\log y)^{t-1}}{(t-1)!}. \tag{7.26}$$

We should mention an additional restriction that we have imposed implicitly upon d : $d > 0$ is sufficiently large so that all poles of $\Phi(s)$ and $\Psi(s)$ lie in the vertical strip $|\mathrm{Re}\, s| < d$. It is also important to note that in the derivation of (7.25) and (7.26) we require both Stirling's formula and the Phragmén-Lindelöf Theorem for a vertical strip. (See the proof of Theorem 2.1 for more details.)

At this juncture we invoke the functional equation (7.5) in (7.25) to obtain

$$g(iy) - b_0 = \frac{\gamma^{-1}}{2\pi i}\int_{(-d)} \Phi(k-s) y^{-s} dy + \hat{p}(y),$$

where $\hat{p}(y)$ is the finite sum on the right-hand side of (7.25). With a change of variables in the integral, we find that

$$\begin{aligned} g(iy) - b_0 &= \frac{\gamma^{-1}y^{-k}}{2\pi i}\int_{(k+d)} \Phi(s) y^{s} ds + \hat{p}(y) \\ &= \frac{\gamma^{-1}y^{-k}}{2\pi i}\int_{(k+d)} \Phi(s) \left(\frac{1}{y}\right)^{-s} ds + \hat{p}(y) \\ &= \gamma^{-1}y^{-k}\{f(i/y) - a_0\} + \hat{p}(y), \end{aligned}$$

by (7.24). This can be rewritten as

$$(\tau/i)^{-k} f(-1/\tau) = \gamma g(\tau) + a_0(\tau/i)^{-k} - \gamma b_0 - \hat{p}(\tau/i), \tag{7.27}$$

valid for $\tau = iy$, $y > 0$. By analytic continuation, (7.27) holds for all τ in

$\mathcal{H}$. Thus we have proved (7.4) with

$$q(\tau) = a_0(\tau/i)^{-k} - \gamma b_0 - \hat{p}(\tau/i),$$

an LPS. This completes the proof of Theorem 7.1. □

Information beyond what is contained in the statement of Theorem 7.1 can be extracted easily from the proof. This is contained in the following.

Remark 7.4. 1. The proof that (B) implies (A) shows immediately that the "period function" $q(\tau)$ in (7.4) can be written as

$$q(\tau) = a_0(\tau/i)^{-k} - \gamma b_0 - \hat{q}(\tau),$$

where

$$\hat{q}(\tau) = \sum_{1\le j\le L'} (\tau/i)^{-\eta_j} \sum_{1\le t\le N'(j)} b'(j,t)\frac{(-1)^{t-1}}{(t-1)!}\log^{t-1}(\tau/i).$$

Note that $\hat{q}(\tau)$ is simply the finite sum in (7.25), with y replaced by τ/i.

2. The proof that (B) implies (A) can be modified by applying the functional equation (7.5) in (7.26) instead of (7.25). This leads to

$$f(\tau) = \gamma(i/\tau)^k g(-1/\tau) + a_0 - \gamma b_0(i/\tau)^k + p(\tau/i),$$

for $\tau = iy$, $y > 0$, where $p(y)$ is the finite sum on the right-hand side of (7.26). Replacing τ by $-1/\tau$ and invoking the identity theorem for analytic functions leads to

$$(\tau/i)^{-k} f(-1/\tau) = \gamma g(\tau) + a_0(\tau/i)^{-k} - \gamma b_0 + (\tau/i)^{-k} p(i/\tau), \qquad (7.28)$$

for $\tau \in \mathcal{H}$. A comparison of (7.28) with (7.27) yields

$$(\tau/i)^{-k} p(i/\tau) = -\hat{p}(\tau/i),$$

and so from the explicit expressions for p, $\hat{p}$ given in (7.25) and (7.26), we conclude that

$$\sum_{1\le j\le L} (\tau/i)^{\delta_j - k} \sum_{1\le t\le N(j)} b(j,t)\frac{1}{(t-1)!}\log^{t-1}(\tau/i)$$
$$= - \sum_{1\le j\le L'} (\tau/i)^{-\eta_j} \sum_{1\le t\le N'(j)} b'(j,t)\frac{(-1)^{t-1}}{(t-1)!}\log^{t-1}(\tau/i).$$

At this point we turn to a description, without detailed proofs, of an application of Theorem 7.1.

Application. This application relates to our Theorem 4.1 ($\lambda > 2$) and Theorem 6.3 ($\lambda = 2$), and especially to Hamburger's Theorem [38], an important result mentioned in Chapter 1 and proved in Chapter 6. (See the Example following Corollary 6.1.) Note that Theorem 4.1 demonstrates the existence of many Dirichlet series of signature (λ, k, γ), as long as $\lambda > 2$, while Theorem 6.3 and Hamburger's Theorem $\left(\text{signature}\left(2, \frac{1}{2}, 1\right)\right)$ both express the paucity of such Dirichlet series for $\lambda = 2$. The application of Theorem 7.1 at hand treats the cases $\lambda = 2$ and $\lambda > 2$.

Hamburger's Theorem [38], [49], [99], [100, pp. 154–156] can be formulated as follows. Put $\Phi(s) = \pi^{-s}\Gamma(s)\varphi(s)$ and assume that $\Phi(s)$ satisfies the three conditions:

1. There exists a polynomial $P(s)$ such that $P(s)\varphi(s)$ is an entire function of finite genus.

2. $\Phi\left(\frac{1}{2} - s\right) = \Phi(s)$, for s in $\mathbb{C}$.

3. Not only $\varphi(s)$, but $\varphi(s/2)$ as well, can be expanded in a Dirichlet series convergent somewhere: $\varphi(s) = \sum b(n)n^{-2s}$. Then, $\varphi(s) = \alpha\zeta(2s)$, where α is a complex constant.

Some fifteen years later, in 1936, Hecke [47], [49] proved an alternative version of Hamburger's Theorem. In the Hecke formulation, the polynomial is specified as $P(s) = s - \frac{1}{2}$ in Condition 1, but Hecke assumes only the expressibility of $\varphi(s)$ itself as a Dirichlet series $\varphi(s) = \sum b(n)n^{-s}$ in Condition 3, not that of $\varphi(s/2)$. Hecke's conclusion is the same: $\varphi(s) = \alpha\zeta(2s)$. Of course, the two proofs differ to some extent. It is important to note here that the Hamburger Theorem we proved in Chapter 6 is the later (that is, the Hecke) formulation.

With these contrasting versions of Hamburger's Theorem at hand, it appears natural to relax both the expressibility of $\varphi(s/2)$ as a Dirichlet series in (3) and the restriction upon the poles of $\varphi(s)$ in Hecke's version, to conjecture that $\varphi(s)$ is uniquely determined by Conditions 1, 2 and the following modification of Condition 3.

3′. Suppose only that $\varphi(s)$ (not $\varphi(s/2)$) can be expanded in a Dirichlet series convergent somewhere: $\varphi(s) = \sum b(n)n^{-s}$.

This conjectured "strong Hamburger's Theorem" has the appealing feature of containing both versions, but unfortunately it fails spectacularly. Indeed [61, Theorem 1] presents the following result.

Abundance principle for Dirichlet series with functional equation. There exist infinitely many linearly independent Dirichlet series $\varphi(s)$ satisfying

Conditions 1, 2, and 3′.

The proofs of the abundance principle and its several generalizations (see [61], [62], [66]) fall into two steps.

Step 1. We translate the question of the existence of the desired Dirichlet series into a question of the existence of many linearly independent exponential series $f(\tau) = \sum_{n=0}^{\infty} a_n e^{\pi i n \tau}$ satisfying

$$(\tau/i)^{-\frac{1}{2}} f(-1/\tau) = f(\tau) + q(\tau),$$

with $q(\tau)$ a LPS. Clearly, this translation is accomplished by Theorem 7.1 with $f = g$, $\lambda_1 = \lambda_2 = 2$, and $\gamma = 1$.

Step 2. We construct, by means of Eichler's generalized Poincaré series [31], infinitely many linearly independent exponential series of the kind described in Step 1.

As stated, the abundance principle stands in sharp contrast to Hamburger's Theorem. The generalizations (there are three) to which we refer above should be compared with Theorem 4.1 and Theorem 6.3. The proof in each case follows the two-step procedure we have already described. The three generalizations are as follows.

1. With $\lambda = 2$, we generalize the principle to arbitrary real k and $\gamma = \pm 1$. This can be contrasted with Theorem 6.3.

2. Again with $\lambda = 2$ and $\gamma = \pm 1$, we restrict k to $k \geq 2$. In this case, we construct $\Phi(s) = \pi^{-s}\Gamma(s)\varphi(s)$ meromorphic in $\mathbb{C}$, with at most a finite number of poles, *at preassigned points and with preassigned principal parts at these points.* This, of course, strengthens the first generalization above considerably in the range $k \geq 2$. One can think of this as a "Mittag-Leffler Theorem" for Dirichlet series with functional equations.

3. This is a generalization to arbitrary $\lambda > 2$, arbitrary real k, and $\gamma = \pm 1$. With $\lambda > 2$ we obtain the stronger Mittag-Leffler version of the abundance principle for all real k. The comparison here is with Theorem 4.1, which gives infinitely many linearly independent solutions of the requisite functional equation, but yields nothing like the Mittag-Leffler result of the generalization.

We now turn our attention to questions of how the sequences $\{a_n\}$, $\{b_n\}$ in Theorem 2.1 may be related. That is to say: suppose $\varphi(s)$, $\psi(s)$ are the Dirichlet series given in Theorem 2.1, and assume that these series are related by the functional equation (2.1). We then wish to determine the extent to which these sequences of coefficients and other relevant parameters can differ. Equivalently, of course, we may formulate this problem instead in terms of the corresponding exponential series $f(\tau)$, $g(\tau)$ satisfying the modular transformation law (i).

We broaden the context of the investigation somewhat by considering the generalization of Theorem 2.1 obtained as a corollary of Theorem 7.1 simply by assuming that $q(\tau) = 0$, where $q(\tau)$ is the log-polynomial sum occurring in (7.4). This leads to a modification of Theorem 2.1 in which the periods λ_1, λ_2 of the exponential series $f(\tau)$, $g(\tau)$ are not necessarily the same. Note that, in light of the proof of Theorem 7.1, the assumption $q(\tau) = 0$ is equivalent to the restriction of the poles of Φ and Ψ to the set $\{0, k\}$.

Below, we state several results that address the issues we have raised here. In proving these, we deviate somewhat from the practice adopted elsewhere in the book, the practice, that is, of presenting self-contained proofs, quoting results that we leave unproved only when these are widely regarded as standard. (Examples are the Cahen-Mellin formula, the Phragmén-Lindelöf theorem, and Stirling's formula, all quoted in the proof of Theorem 2.1.) For, strict adherence to this practice in the proofs to be given in the remainder of the chapter would lead us far afield and create imbalance in the exposition. We trust that the effects of this change in practice will be mitigated in part by the specific references to the literature we supply for the theorems quoted without proof.

The theorems we present address the three questions:

(i) Which parameters λ_1, λ_2 can occur in nontrivial pairs Φ, Ψ, defined as in (7.3), satisfying the functional equation (7.5), and with poles restricted to the set $\{0, k\}$? (Recall that the latter condition is equivalent to our assumption that $q(\tau) = 0$ in (7.4).)

(ii) With λ_1, λ_2 fixed, consistent with the answer to (i), to what extent can Φ and Ψ differ?

(iii) How is the weight parameter k in (7.5) related to λ_1, λ_2?

In partial answer, we present Theorems 7.2, 7.3, and 7.4, below. Before we state these, we alter slightly the notation of Theorem 7.1, incorporating the factor γ into the function g (effectively putting $\gamma = 1$). Thus, for example, (7.5) becomes $\Phi(k - s) = \Psi(s)$. Henceforth, the factor γ will be reserved for the special case in which $g = \gamma f$ (that is, $\Psi = \gamma\Phi$). Of necessity, then, $\gamma = \pm 1$.

Theorem 7.2. *Suppose Φ, Ψ are defined as in (7.3) with coefficient sequences of polynomial growth. Assume further that Φ and Ψ are noncon-*

stant, related by (7.5) (i.e., $\Phi(k-s) = \Psi(s)$) and with poles (if any) confined to the set $\{0, k\}$. Then, either $\lambda_1\lambda_2 = 4\cos^2 \pi/q$, with $q \in \mathbb{Z}$, $q \geq 3$, or $\lambda_1\lambda_2 \geq 4$.

Theorem 7.3. *Assume that λ_1, $\lambda_2 > 0$, with $\lambda_1\lambda_2 = 4\cos^2 \pi/q$, where q is an odd integer ≥ 3. Further, assume that Φ and Ψ are as in Theorem 7.2. Then,*

(i) $k(1 - q/2) \in \mathbb{Z}$;

(ii) $\Psi(s) = e^{\pi i k(1-q/2)}(\lambda_1/\lambda_2)^{k/2-s}\Phi(s)$.

Theorem 7.4. *Assume that λ_1, λ_2 are as in Theorem 7.3, except that now q is even. Let Φ, Ψ be as in Theorems 7.2 and 7.3. Then, $k(1-q/2) \in 2\mathbb{Z}$.*

Remark 7.5. 1. There are results that address the question to what extent Theorems 7.2, 7.3, and 7.4 are best possible. These will be stated, and their proofs sketched, after we have completed the proof of Theorem 7.4.

2. Note that the expression $e^{\pi i k(1-q/2)} = +1$ or -1, according as $k(1 - q/2)$ is even or odd.

The proofs of these theorems require a modest amount of information concerning "multiplier systems" on $G(\lambda)$ and on a certain subgroup of index 2 in $G(\lambda)$. Thus we now present a formal definition and a brief discussion of how multiplier systems arise within the context of the Hecke correspondence and Bochner's generalization. (Multiplier systems have been mentioned earlier in the book, but only in passing and indirectly. They are, of course, related closely to the "multipliers" of Definition 5.2.)

Definition 7.2. Suppose that k is a real number and Γ is a discrete subgroup of $SL(2, \mathbb{R})$. A *multiplier system* (MS) v *of weight* k *on* Γ is a function from Γ onto $\mathbb{C}$, such that

$$\text{(a)} \qquad |v(M)| = 1, \qquad \text{for all } M \in \Gamma;$$

$$\begin{aligned}\text{(b)} \qquad & v(M_1M_2)(-i(c_3\tau + d_3))^k \\ & = v(M_1)v(M_2)(-i(c_1M_2\tau + d_1))^k(-i(c_2\tau + d_2))^k, \\ & \qquad \text{for all } M_1,\ M_2 \in \Gamma. \end{aligned} \tag{7.29}$$

In (b), $M_1M_2 = M_3$ and $M_j = \begin{pmatrix} * & * \\ c_j & d_j \end{pmatrix}$, $1 \leq j \leq 3$. The stipulation (7.29) is usually called the *consistency condition for multiplier systems* for MS's with respect to (Γ, k).

Remark 7.6. 1. In (b), as throughout, we adhere to the argument convention (1.4).

2. If $k \in \mathbb{Z}$, (b) reduces to $v(M_1M_2) = e^{-\pi ik/2}v(M_1)v(M_2)$. Thus, if $k \equiv 0 \pmod 4$, v is a multiplicative character on the matrix group Γ.

3. As an immediate consequence of (b) (and $v(M) \neq 0$), we find that $v(I) = e^{\pi ik/2}$ and $v(-I) = \pm e^{-\pi ik/2}$.

The motivation for the condition (7.29) is the characteristic transformation law of an automorphic form F of weight k and MS v, with respect to Γ:

$$F(M\tau) = v(M)(-i(c\tau + d))^k F(\tau), \qquad \tau \in \mathcal{H}, \tag{7.30}$$

with $k \in \mathbb{R}$, $v(M) \in \mathbb{C}$, and $|v(M)| = 1$, for all $M = \left(\begin{smallmatrix} * & * \\ c & d \end{smallmatrix}\right) \in \Gamma$. Indeed, the existence of $F \neq 0$ satisfying (7.30) implies directly that v satisfies (7.29). Note that if F is meromorphic in $\mathcal{H}$, it follows from (7.30) and the fact $|v(M)| = 1$ that $v(M)$ is independent of τ. It is customary to call F an *automorphic form of weight k and* MS *v on Γ*, provided that F satisfies (7.30) and certain growth conditions at the real points of the boundary of a fundamental region for Γ. (Compare this with the more restrictive definition of $M(\lambda, k, \gamma)$ at the end of Chapter 2.)

The existence of $F \neq 0$ satisfying (7.2) implies that

$$v(-I) = +e^{-\pi ik/2}. \tag{7.31}$$

For obvious reasons, (7.31) is usually called the *nontriviality condition for multiplier systems* for MS's in weight k. Here we consider only those MS's satisfying (7.31).

Before starting the proofs of our theorems, we elucidate the connection of MS's with our previous work—in particular, with the Hecke correspondence theorem and the Bochner generalization. In the former, Theorem 2.1, we are given two exponential series

$$f(\tau) = \sum_{n=0}^{\infty} a_n e^{2\pi in\tau/\lambda}, \qquad g(\tau) = \sum_{n=0}^{\infty} b_n e^{2\pi in\tau/\lambda},$$

both convergent for $\tau \in \mathcal{H}$. We further assume that the pair f, g satisfies condition (i):

$$f(\tau) = (\tau/i)^{-k} g(-1/\tau), \qquad k \in \mathbb{R},\ \tau \in \mathcal{H}. \tag{7.32}$$

We make the simplifying assumption that $g = \gamma f$, so that $f \in M(\lambda, k, \gamma)$ and $\gamma = \pm 1$. (See Definition 2.2 of Chapter 2.) Then we have the two transformation laws for f:

$$f(\tau+\lambda)=f(\tau), \qquad f(-1/\tau)=\gamma(\tau/i)^k f(\tau). \tag{7.33}$$

Recall the notation $S_\lambda\tau=\tau+\lambda$, $T\tau=-1/\tau$, and $G(\lambda)=\langle S_\lambda, T\rangle$. We show that (7.33) can be invoked to generate a MS v on $G(\lambda)$ of weight k, at the same time proving that $f(\tau)$ satisfies (7.30) with MS v for all $M\in G(\lambda)$. We proceed by expressing $M\in G(\lambda)$ as a word in S_λ and T:

$$M=T^{\epsilon_1}S_\lambda^{\ell_1}TS_\lambda^{\ell_2}\cdots S_\lambda^{\ell_n}T^{\epsilon_2}, \tag{7.34}$$

with $\ell_j\in\mathbb{Z}$, $\ell_j\neq 0$, $\epsilon_1=0$, 1, 2, or 3 and $\epsilon_2=0$ or 1. This representation of M as a word in S_λ, T is not unique since there are nontrivial relations in $G(\lambda)$, for example, $T^2=-I$. However, as we shall observe, this has no bearing upon the proof we are about to give, since $v(M)$, as we generate it, is uniquely determined by $f(\tau)$ and M, in light of (7.30), with $F=f$.

The procedure is iterative. Define $v(S_\lambda)=1$, $v(T)=\gamma$, as dictated by (7.33). For general $M\in G(\lambda)$, apply the expression (7.34) and make repeated use of the consistency condition (7.29) in the form

$$v(M_1M_2)=v(M_1)v(M_2)\frac{(-i(c_1M_2\tau+d_1))^k(-i(c_2\tau+d_2))^k}{(-i(c_3\tau+d_3))^k}. \tag{7.35}$$

Note that, since $|v(S_\lambda)|=|v(T)|=1$ and the ratio on the right-hand side of (7.30) has absolute value 1 as well, it follows that $|v(M_1M_2)|=1$, as required. This approach yields (7.30) with $F=f$, MS v, and $\Gamma=G(\lambda)$. Notwithstanding the nonuniqueness of the expression (7.34), the uniqueness of $v(M)$ follows directly from (7.30), since

$$v(M)=\frac{f(M\tau)}{f(\tau)}(-i(c\tau+d))^{-k}.$$

Note that (7.29) and (7.31) also follow immediately from (7.30).

Of course, this procedure is applicable to discrete groups other than $G(\lambda)$ as well. Indeed, in the proofs of Theorems 7.2, 7.3, and 7.4 we shall have occasion to apply the method to the group $K(\lambda)=\langle S_\lambda, TS_\lambda T\rangle$, a subgroup of $G(\lambda)$, and to a generalized version of $K(\lambda)$. In the end, the proofs of all three theorems come down to establishing the existence of a nontrivial function on $\mathcal{H}$, satisfying (7.30) with an appropriate MS on some group related to $G(\lambda)$.

Remark 7.7. It is useful to know that the procedure outlined above works equally well if $f(\tau)$, $g(\tau)$ are related by a transformation law of the form (7.4) with a log-polynomial sum $q(\tau) \neq 0$. However, the proof is of necessity more complicated. We do not give it here.

Proof of Theorem 7.2. Under the assumptions of Theorem 7.2, Theorem 7.1 implies that the pair $f(\tau)$, $g(\tau)$ satisfies (7.4) (with $\gamma = 1$), where $f(\tau)$, $g(\tau)$ are given by (7.2) and are related to $\Phi(s)$, $\Psi(s)$, respectively, by (7.3). Furthermore, an examination of the proof of Theorem 7.1 shows that, under restriction of the poles of $\Phi(s)$ and $\Psi(s)$ to the set $\{0, k\}$, $q(\tau) = 0$ in (7.4). Thus, in this case we have the stricter form of (7.4),

$$(\tau/i)^{-k} f(-1/\tau) = g(\tau), \qquad \tau \in \mathcal{H}. \tag{7.36}$$

Since $g(\tau + \lambda_2) = g(\tau)$, it follows that

$$\left(\frac{\tau + \lambda_2}{i}\right)^{-k} f\left(\frac{-1}{\tau + \lambda_2}\right) = g(\tau).$$

Replacing τ by $-1/\tau$ and applying (7.36) once again, we find that

$$\left(\frac{\lambda_2 \tau - 1}{i\tau}\right)^{-k} f\left(\frac{-\tau}{\lambda_2 \tau - 1}\right) = \left(\frac{-1}{i\tau}\right)^{-k} f(\tau),$$

that is to say,

$$f\left(\frac{-\tau}{\lambda_2 \tau - 1}\right) = \left(\frac{\lambda_2 \tau - 1}{i\tau}\right)^{k} \left(\frac{-1}{i\tau}\right)^{-k} f(\tau).$$

Since

$$\left| \left(\frac{\lambda_2 \tau - 1}{i\tau}\right)^{-k} \left(\frac{-1}{i\tau}\right)^{k} \right| = |\lambda_2 \tau - 1|^k,$$

it follows that

$$f\left(\frac{-\tau}{\lambda_2 \tau - 1}\right) = \epsilon \left(\frac{\lambda_2 \tau - 1}{i}\right)^{k} f(\tau), \tag{7.37}$$

with $|\epsilon| = 1$. As before, the condition $|\epsilon| = 1$ implies that

$$\epsilon = \left(\frac{\lambda_2 \tau - 1}{i}\right)^{-k} \frac{f\left(\frac{-\tau}{\lambda_2 \tau - 1}\right)}{f(\tau)}$$

is independent of $\tau \in \mathcal{H}$. From the definition of $f(\tau)$, $f(\tau + \lambda_1) = f(\tau)$, and so by the procedure outlined above (for the group $G(\lambda) = \langle S_\lambda, T \rangle$), we find that $f(\tau)$ satisfies (7.30) for the group $\Gamma = K(\lambda_1, \lambda_2) = \langle S_{\lambda_1}, TS_{\lambda_2}T\rangle$, with $v = v_{\lambda_1,\lambda_2,k}$, a MS of weight k on $K(\lambda_1, \lambda_2)$.

By assumption, $\Phi(s) \not\equiv 0$, so $f(\tau)$ is not constant. Since $K(\lambda_1, \lambda_2)$ contains translations, Corollary 14 of [59] implies that $K(\lambda_1, \lambda_2)$ is discrete. Thus, by [65, Theorem 1], we conclude that either $\lambda_1\lambda_2 = 4\cos^2 \pi/q$, with $q \in \mathbb{Z}$, $q \geq 3$, or $\lambda_1\lambda_2 \geq 4$. This competes the proof of Theorem 7.2. □

In Theorems 7.3 and 7.4 we deal only with the first of the two alternatives in the conclusion of Theorem 7.2. That is, we treat the case $\lambda_1\lambda_2 = 4\cos^2 \pi/q$, with $q \in \mathbb{Z}$, $q \geq 3$. In Theorem 7.3 we assume in addition that q is odd. In the proof of the latter, we require a preliminary result.

Proposition 7.2. *For $\lambda = 2\cos \pi/q$, $q \in \mathbb{Z}$, $q \geq 3$, we have the following relations in $G(\lambda)$:*

$$T^2 = (TS_\lambda)^q = -I, \tag{7.38}$$

with $T = \left(\begin{smallmatrix} 0 & -1 \\ 1 & 0 \end{smallmatrix}\right)$ and $S_\lambda = \left(\begin{smallmatrix} 1 & \lambda \\ 0 & 1 \end{smallmatrix}\right)$.

Proof. The first relation, $T^2 = -I$, is clear. The second was proved in Chapter 5, but stated in a slightly different form. Recall that

$$T_1\tau = \frac{\tau}{|\tau|^2}, \qquad T_3\tau = -(\bar{\tau} + \lambda),$$

as defined in the proof of Lemma 3.1; by a simple calculation, then, $T_1T_3 = TS_\lambda$. In the proof of Theorem 5.2, we observed that

$$(T_1T_3)^n = \frac{1}{\sin(\pi p/q)} \begin{bmatrix} \sin\{\pi p(1-n)/q\} & -\sin(\pi pn/q) \\ \sin(\pi pn/q) & \sin\{\pi p(1+n)/q\} \end{bmatrix},$$

with $\lambda = 2\cos(\pi p/q)$, $(p, q) = 1$, $q \geq 3$, $p \geq 2$, and $n \in \mathbb{Z}^+$. By (3.1), this holds for $p = 1$ as well. Putting $p = 1$ and $n = q$, we deduce that

$$\begin{aligned} (T_1T_3)^n &= \frac{1}{\sin(\pi/q)} \begin{bmatrix} \sin\{\pi(1-q)/q\} & -\sin\pi \\ \sin\pi & -\sin(\pi/q) \end{bmatrix} \\ &= \frac{1}{\sin(\pi/q)} \begin{bmatrix} -\sin(\pi/q) & 0 \\ 0 & -\sin(\pi/q) \end{bmatrix} = -I, \end{aligned}$$

as asserted. □

Proof of Theorem 7.3. We begin with the proof of Theorem 7.3 for the special case $\lambda_1 = \lambda_2$. Thus, we may write $\lambda = \lambda_1 = \lambda_2 = 2\cos \pi/q$, with q odd and $q \geq 3$. Just as in the proof of Theorem 7.2, we note that $f(\tau)$

and $g(\tau)$ are related by the transformation law (7.36), so we obtain a MS $v_{\lambda,k}$ of weight k on $K(\lambda)$. (Note the abbreviated notations $v_{\lambda,k} = v_{\lambda,\lambda,k}$ and $K(\lambda) = K(\lambda,\lambda)$.) Again, $f(\tau)$ satisfies (7.30) for the group $K(\lambda)$ with $v = v_{\lambda,k}$. Since, by assumption, $f(\tau)$ is nonconstant, $v_{\lambda,k}$ satisfies the nontriviality condition (7.31).

Since q is odd, the second relation in (7.38) can be used to express the inversion T in $G(\lambda)$ as a word in S_λ and $TS_\lambda T$ of length q, namely,

$$(TS_\lambda T)S_\lambda(TS_\lambda T)S_\lambda \cdots (TS_\lambda T) = -T. \tag{7.39}$$

(In (7.39) there are $(q+1)/2$ factors $TS_\lambda T$ separated by $(q-1)/2$ factors S_λ.) In particular, this shows that $G(\lambda) = K(\lambda) = \langle S_\lambda, TS_\lambda T\rangle$. Hence, $f(\tau)$ satisfies (7.30) for all M in $G(\lambda)$, including $M = T$, i.e.,

$$f(-1/\tau) = v_{\lambda,k}(T)(-i\tau)^k f(\tau),$$

or

$$(\tau/i)^{-k} f(-1/\tau) = v_{\lambda,k}(T) f(\tau). \tag{7.40}$$

A comparison with (7.36) implies that $g(\tau) = v_{\lambda,k}(T)f(\tau)$, so that $\Psi(s) = v_{\lambda,k}(T)\Phi(s)$. Except for the specific determination of the factor $v_{\lambda,k}(T)$ on the right-hand side in terms of k and q, this proves part (ii) of the theorem in the special case $\lambda_1 = \lambda_2$.

We invoke (5.18) from Theorem 5.6, which yields $k = 4m/(q-2) + 1 - v_{\lambda,k}(T)$, with $m \in \mathbb{Z}^+$, since $v_{\lambda,k}(T) = \gamma$, in the notation of Definition 2.2 and Theorem 5.6. Furthermore, it is easy to show that $v_{\lambda,k}(T) = \pm 1$, either directly from

$$\Psi(k-s) = \Phi(s) = v_{\lambda,k}(T)^{-1}\Psi(s)$$

or by applying (7.29). This implies that

$$k\left(1 - \frac{q}{2}\right) = -2m + \left(\frac{\gamma - 1}{2}\right)(q-2) \in \mathbb{Z},$$

and so we have proved part (i) of the theorem. Since q is odd here, note that $k(1-q/2) \in 2\mathbb{Z}$ if and only if $\gamma = 1$. From this it follows directly that $e^{\pi i k(1-q/2)} = e^{\pi i (q-2)(\gamma-1)/2} = \gamma = v_{\lambda,k}(T)$, since $\gamma = \pm 1$. This completes the proof of Theorem 7.3 in the case $\lambda_1 = \lambda_2$.

Extension of the proof to the general case. At this point we lift the restriction $\lambda_1 = \lambda_2$, assuming only that λ_1, $\lambda_2 > 0$ and $\lambda_1\lambda_2 = 4\cos^2 \pi/q$, with $q \in \mathbb{Z}$, $q \geq 3$, and q odd. The assumptions, once again, are that $\Phi(s)$,

$\Psi(s)$ are nonconstant, defined as in Theorem 7.1, related by the functional equation $\Psi(k-s)=\Phi(s)$ ((7.5) with $\gamma=1$), and meromorphic in $\mathbb{C}$, with poles confined to the set $\{0,k\}$.

Let $\lambda=\sqrt{\lambda_1\lambda_2}=2\cos\pi/q$ and define

$$\widehat{\Phi}(s)=\lambda_1^{k/2}\left(\sqrt{\frac{\lambda_2}{\lambda_1}}\right)^s\Phi(s),$$

$$\widehat{\Psi}(s)=\lambda_2^{k/2}\left(\sqrt{\frac{\lambda_1}{\lambda_2}}\right)^s\Psi(s).$$

We note that

$$\widehat{\Phi}(s)=\left(\frac{2\pi}{\lambda}\right)^{-s}\sum_{n=1}^{\infty}\widehat{a}_n n^{-s},$$

$$\widehat{\Psi}(s)=\left(\frac{2\pi}{\lambda}\right)^{-s}\sum_{n=1}^{\infty}\widehat{b}_n n^{-s},$$

with $\widehat{a}_n=\lambda_1^{k/2}a_n$, $\widehat{b}_n=\lambda_2^{k/2}b_n$, for all n. A simple calculation employing the functional equation $\Psi(k-s)=\Phi(s)$ shows that $\widehat{\Psi}(k-s)=\widehat{\Phi}(s)$. The theorem for the special case $\lambda_1=\lambda_2$ now implies that $k(1-q/2)\in\mathbb{Z}$ and $\widehat{\Psi}(s)=v_{\lambda,k}(T)^{-1}\widehat{\Phi}(s)$, that is,

$$\Psi(s)=v_{\lambda,k}(T)^{-1}\left(\frac{\lambda_1}{\lambda_2}\right)^{k/2-s}\Phi(s).$$

As before, $v_{\lambda,k}(T)=e^{\pi i-k(1-q/2)}$. This completes the proof of Theorem 7.3. □

Proof of Theorem 7.4. The simple considerations of the previous paragraph permit us to assume at the outset that $\lambda_1=\lambda_2=\lambda$. As in the proof of Theorem 7.3, the hypotheses of Theorem 7.4 imply that $f(\tau)$ and $g(\tau)$ are related by the transformation formula (7.36) with the consequence that once again $f(\tau)$ satisfies (7.30) for the group $\Gamma=K(\lambda)=\langle S_\lambda, TS_\lambda T\rangle$, with a MS $v=\tilde{v}_{\lambda,k}$ of weight k on $K(\lambda)$. Since q is even in this case, it is *not* true that $K(\lambda)=G(\lambda)$, but, in fact, $[G(\lambda):K(\lambda)]=2$. Thus, the relations (7.38) do not imply that $f(\tau)$ satisfies (7.30) for $M=T$, and we cannot prove anything like Theorem 7.3 (b).

Nevertheless, the MS $\tilde{v}_{\lambda,k}$ does extend to $G(\lambda)$, and this will suffice to obtain the restriction upon the weight k that we claim. To show this, we revisit the transformation formula (7.36), i.e.,

$$(\tau/i)^{-k} f(-1/\tau) = g(\tau), \qquad \tau \in \mathcal{H}.$$

Replacing τ by $-1/\tau$ and rearranging yields

$$(\tau/i)^{-k} g(-1/\tau) = f(\tau), \qquad \tau \in \mathcal{H}.$$

Letting $h_1(\tau) = f(\tau) + g(\tau)$ and $h_2(\tau) = f(\tau) - g(\tau)$, we find that

$$(\tau/i)^{-k} h_1(-1/\tau) = g(\tau) + f(\tau) = h_1(\tau)$$

and

$$(\tau/i)^{-k} h_2(-1/\tau) = g(\tau) - f(\tau) = -h_2(\tau).$$

Since $f(\tau)$, $g(\tau)$ are nonconstant, it follows that $h_1(\tau) = h_2(\tau) = 0$ cannot occur, so there is a MS $v_{\lambda,k}$ of weight k on all of $G(\lambda)$, satisfying the nontriviality condition, which restricts to $\tilde{v}_{\lambda,k}$ on the subgroup $K(\lambda)$. (If $h_1(\tau) \neq 0$, we have $v_{\lambda,k}(T) = +1$; if $h_2(\tau) \neq 0$, we have $v_{\lambda,k}(T) = -1$; if both are $\neq 0$, we have two distinct MS's on $G(\lambda)$ that restrict to $\tilde{v}_{\lambda,k}$ on $K(\lambda)$.)

As in the proof of Theorem 7.3, we may now apply (5.18) of Theorem 5.6 to conclude that $k = 4m/(q-2) + 1 - v_{\lambda,k}(T)$. Thus,

$$k\left(1 - \frac{q}{2}\right) = -2m + \left\{\frac{v_{\lambda,k}(T) - 1}{2}\right\}(q-2)$$

is an even integer, since q is even and $v_{\lambda,k}(T) = \pm 1$. This completes the proof of Theorem 7.4. □

Theorems 7.2, 7.3, and 7.4 may be regarded as nonexistence results, since they impose restrictions upon the complex constants that occur in Theorem 7.1, subject to the restriction $q(\tau) = 0$ in (7.4). The article [65] describes the values of λ_1, $\lambda_2 > 0$ that yield discrete $K(\lambda_1, \lambda_2) = \langle S_{\lambda_1}, TS_{\lambda_2}T\rangle$. They are:

(a) $\lambda_1\lambda_2 = 4\cos^2(\pi/q)$, $q \in \mathbb{Z}$, $q \geq 3$, with q odd;
(b) $\lambda_1\lambda_2 = 4\cos^2(\pi/q)$, $q \in \mathbb{Z}$, $q \geq 3$, with q even;
(c) $\lambda_1\lambda_2 = 4$;
(d) $\lambda_1\lambda_2 > 4$.

A glance at Theorems 7.3 and 7.4 shows that in case (a) the results are far stronger than in (b). In cases (c) and (d) there are no results at all relating the functions and parameters that occur, explicitly or implicitly, in the functional equation (7.5) (with $\gamma = 1$). In Theorem 7.5 we formulate existence results showing that, in fact, the nonexistence statements of Theorems 7.3 and 7.4 cannot be extended further. Furthermore, it follows directly from Theorem 7.5 that Theorem 7.2 is best possible as well.

Theorem 7.5. *Assume $\lambda_1, \lambda_2 > 0$.*

(a) *Let $\lambda_1\lambda_2 = 4\cos^2(\pi/q)$, where q is an odd integer ≥ 3. Suppose that $k(1-q/2) \in \mathbb{Z}$, and put $\gamma = 1$ if $k(1-q/2)$ is even and $\gamma = -1$ if $k(1-q/2)$ is odd. Assume that*

$$1 + \left[k\left(\frac{1}{4} - \frac{1}{2}q\right) + \frac{\gamma - 1}{4}\right] > 0.$$

($k \geq 0$ suffices to ensure this when $\gamma = 1$; when $\gamma = -1$, $k \geq 2q/(q-2)$ is sufficient.) Then there exists nontrivial

$$\Phi(s) = \left(\frac{2\pi}{\lambda_1}\right)^{-s} \Gamma(s)\varphi(s), \qquad \varphi(s) = \sum a_n n^{-s},$$

with $a_n = O(n^\rho)$, $n \to \infty$, $\rho > 0$, such that:

(i) *$\Phi(s)$ has a meromorphic continuation to the entire s-plane, with poles restricted to the set $\{0, k\}$;*

(ii) *$\Phi(s)$ is bounded in lacunary vertical strips (see* (B) *of Theorem 7.1 for the definition);*

(iii) *$\Phi(k-s) = e^{\pi i k(1/q/2)}(\lambda_1/\lambda_2)^{k/2-s}\Phi(s)$.*

If we assume that k satisfies the stronger inequality $k \geq 4q/(q-2)$ in the case $\gamma = 1$, and $k \geq 6q/(q-2)$ in the case $\gamma = -1$, then $\Phi(s)$ can be chosen so that (i) *is strengthened to:*

(iv) *$\Phi(s)$ is an entire function of s.*

(b) *Suppose $\lambda_1\lambda_2 = 4\cos^2(\pi/q)$, where q is an even integer ≥ 3. Let $k(1-q/2) \in 2\mathbb{Z}$ and $\gamma = \pm 1$. As in* (a), *assume that*

$$1 + \left[k\left(\frac{1}{4} - \frac{1}{2q}\right) + \frac{\gamma - 1}{4}\right] > 0,$$

but now for $\gamma = \pm 1$ simultaneously. (These two inequalities reduce to the single one $k \geq 2q/(q-2)$.) Then there exist nontrivial

$$\Phi(s) = \left(\frac{2\pi}{\lambda_1}\right)^{-s} \Gamma(s) \sum a_n n^{-s},$$

$$\Psi(s) = \left(\frac{2\pi}{\lambda_2}\right)^{-s} \Gamma(s) \sum b_n n^{-s},$$

with both $\{a_n\}$ *and* $\{b_n\}$ *of polynomial growth:* $a_n = O(n^\rho)$, $b_n = O(n^\rho)$, $n \to \infty$, $\rho > 0$, *and such that* $\Phi(s)$, $\Psi(s)$ *both satisfy conditions* (i) *and* (ii) *above, and*

(v) $\Psi(k-s) = \Phi(s)$.

Furthermore, $\{a_n\}$ *and* $\{b_n\}$ *can be chosen so that* $\lambda_1^{-s}\Phi(s)$ *and* $\lambda_2^{-s}\Psi(s)$ *are linearly independent over* $\mathbb{C}$. *If we assume that* $k \geq 6q/(q-2)$, *then* (i) *can be strengthened, as in* (a), *to the assertion* (iv) *for* $\Phi(s)$ *and* $\Psi(s)$ *suitably chosen.*

(c) *Suppose* $\lambda_1\lambda_2 = 4$ *and* $\gamma = \pm 1$. *Assume that* $1 + [(k+\gamma-1)/4] > 0$ *for both* $\gamma = +1$ *and* $\gamma = -1$, *but nothing further about* k. (*The two inequalities taken together are equivalent to* $k \geq 2$.) *The conclusion is the same as in* (b), *provided the last sentence in case* (b) *is omitted.*

(d) *Let* $\lambda_1\lambda_2 > 4$ *and* $k \in \mathbb{R}$, *with* k *arbitrary. The conclusion here is the same as in* (c), *but with the additional feature that for each* k *there are infinitely many linearly independent choices for each of* $\Phi(s)$ *and* $\Psi(s)$.

Remark 7.8. 1. Note that the right-hand side of (iii) has the form

$$\gamma \left(\frac{2\pi}{\lambda_2}\right)^{-s} \Gamma(s) \left(\frac{\lambda_1}{\lambda_2}\right)^{k/2} \varphi(s) = \pm \left(\frac{2\pi}{\lambda_2}\right)^{-s} \Gamma(s) \left(\frac{\lambda_1}{\lambda_2}\right)^{k/2} \varphi(s).$$

2. In part (a), $\Phi(s)$ can be replaced by

$$\Psi(s) = \left(\frac{2\pi}{\lambda_2}\right)^{-s} \Gamma(s) \sum b_n n^{-s},$$

with λ_1 and λ_2 reversing roles in (iii).

3. In case (c) there is a further result analogous to the final statements in (a) and (b), but the proof of this depends on knowing that $\dim C^0(2,k,\gamma) > 0$ for k "large." This is, of course, well known, but we have not discussed it here. (It can be proved from Theorem 6.3 and the introduction of Eisenstein series more general than those of Definitions 5.3 and 5.4.) What can be shown is this: if $k > 2$, then both $\Phi(s)$ and $\Psi(s)$ can be chosen continuable to entire functions of s. The same addition can be made to (d), for $k \geq 2$.

Proof of Theorem 7.5. The proofs of the results included here are based upon the dimensionality results, Theorems 5.6 and 5.8 ($\lambda < 2$), Theorem 6.3 ($\lambda = 2$), and Theorem 4.1 ($\lambda > 2$). They also involve Theorem 2.1 and Corollary 5.3(a) in an essential way.

(a) $\lambda_1\lambda_2 = 4\cos^2(\pi/q)$, where q is an odd integer ≥ 3. The existence proof in this case, and the following cases as well, begins with the special situation in which both λ_1 and λ_2 are replaced by $\lambda = \sqrt{\lambda_1 \lambda_2}$. In this case, then, $\lambda = 2\cos(\pi/q)$. Suppose $k \in \mathbb{R}$ and $k(1 - q/2) \in \mathbb{Z}$. By Theorem 5.6 and the restrictions on k,

$$\dim M(\lambda, k, e^{\pi ik(1-q/2)}) > 0.$$

(Since $k(1 - q/2) \in \mathbb{Z}$, it follows that $\gamma = e^{\pi ik(1-q/2)} = \pm 1$.) But, by Corollary 5.3(a), $M_0(\lambda, k, \gamma) = M(\lambda, k, \gamma)$, so

$$\dim M_0(\lambda, k, e^{\pi ik(1-q/2)}) > 0.$$

Theorem 2.1 then implies the existence of a nontrivial Dirichlet series $\hat{\varphi}(s)$, convergent in some right half-plane, such that $\hat{\Phi}(s) = \left(\frac{2\pi}{\lambda}\right)^{-s} \Gamma(s)\hat{\varphi}(s)$ satisfies (i), (ii) and the simplified version of (iii) obtained by putting $\lambda_1 = \lambda_2$ (see (2.1)):

$$\hat{\Phi}(k - s) = e^{\pi ik(1-q/2)} \hat{\Phi}(s).$$

The extension to general λ_1, λ_2 (still subject to the condition imposed at the outset, of course) can be carried out by reference to the extension of the proof of Theorem 7.3 to the general case. Simply let

$$\Phi(s) = \lambda_1^{-k/2} \left(\frac{\lambda_2}{\lambda_1}\right)^{s/2} \hat{\Phi}(s);$$

this fulfills the required conditions.

Next, by Theorem 5.8, $\dim C^0(\lambda, k, \gamma) > 0$, provided $k \geq 4q/(q-2)$, for $\gamma = 1$, and $k \geq 6q/(q-2)$, for $\gamma = -1$. (Thus, for example, $k \geq 12$ suffices in the first case and $k \geq 18$ suffices in the second.) By Theorem 2.1, then, these stricter inequalities (k still subject to the condition $k(1-q/2) \in \mathbb{Z}$) guarantee the existence of entire $\Phi(s)$ satisfying the conditions in the conclusion of Theorem 7.5(a). This completes the proof of part (a).

(b) $\lambda_1 \lambda_2 = 4\cos^2(\pi/q)$, where q is an even integer ≥ 3. As in (a), we may assume from the outset that both λ_1 and λ_2 are replaced by $\lambda = 2\cos(\pi/q)$.

Suppose $k \in \mathbb{R}$ is such that $k(1 - q/2) \in 2\mathbb{Z}$ and $\gamma = \pm 1$. Under the further assumption $k \geq 2q/(q-2)$, Theorem 5.6 and Corollary 5.3(a) together imply that $\dim M_0(\lambda, k, \gamma) > 0$ for both $\gamma = +1$ and $\gamma = -1$. Theorem 2.1 then implies the existence of nontrivial Dirichlet series $\hat{\varphi}(s)$, $\tilde{\varphi}(s)$, convergent in some right half-plane, such that both

$$\hat{\Phi}(s) = \left(\frac{2\pi}{\lambda}\right)^{-s} \Gamma(s)\hat{\varphi}(s) \quad \text{and} \quad \tilde{\Phi}(s) = \left(\frac{2\pi}{\lambda}\right)^{-s} \Gamma(s)\tilde{\varphi}(s)$$

satisfy (i) and (ii), and also

$$\hat{\Phi}(k-s) = \hat{\Phi}(s) \quad \text{and} \quad \tilde{\Phi}(k-s) = -\tilde{\Phi}(s). \tag{7.41}$$

Let

$$\Phi^+(s) = \lambda_1^{-k/2}\left(\sqrt{\lambda_1/\lambda_2}\right)^s \hat{\Phi}(s) \quad \text{and} \quad \Psi^+(s) = \lambda_2^{-k/2}\left(\sqrt{\lambda_2/\lambda_1}\right)^s \hat{\Phi}(s).$$

It follows that

$$\Phi^+(s) = \left(\frac{2\pi}{\lambda_1}\right)^{-s} \Gamma(s)\lambda_1^{-k/2}\hat{\varphi}(s) \quad \text{and} \quad \Psi^+(s) = \left(\frac{2\pi}{\lambda_2}\right)^{-s} \Gamma(s)\lambda_2^{-k/2}\hat{\varphi}(s),$$

and from (7.41) we infer that $\Psi^+(k-s) = \Phi^+(s)$. Similarly, defining

$$\Phi^-(s) = \lambda_1^{-k/2}\left(\sqrt{\lambda_1/\lambda_2}\right)^s \tilde{\Phi}(s) \quad \text{and} \quad \Psi^-(s) = -\lambda_2^{-k/2}\left(\sqrt{\lambda_2/\lambda_1}\right)^s \tilde{\Phi}(s),$$

we find that

$$\Phi^-(s) = \left(\frac{2\pi}{\lambda_1}\right)^{-s} \Gamma(s)\lambda_1^{-k/2}\tilde{\varphi}(s), \text{ and } \Psi^-(s) = \left(\frac{2\pi}{\lambda_2}\right)^{-s} \Gamma(s)(-\lambda_2^{-k/2})\tilde{\varphi}(s),$$

and that $\Psi^-(k-s) = \Phi^-(s)$.

Now, the pair $\Phi^+(s)$, $\Psi^+(s)$, satisfies (i), (ii), and (v), as does the pair $\Phi^-(s)$, $\Psi^-(s)$. However, neither pair satisfies the linear independence condition described in the statement of (b) (penultimate sentence), since

$$\lambda_1^{k/2}(\lambda_1^{-s}\Phi^+(s)) = \lambda_2^{k/2}(\lambda_2^{-s}\Psi^+(s)) = (\sqrt{\lambda_1\lambda_2})^{-s}\hat{\Phi}(s),$$

and

$$\lambda_1^{k/2}(\lambda_1^{-s}\Phi^-(s)) = -\lambda_2^{k/2}(\lambda_2^{-s}\Psi^-(s)) = (\sqrt{\lambda_1\lambda_2})^{-s}\tilde{\Phi}(s).$$

In order to achieve the desired linear independence, we define

$$\Phi(s) = \Phi^+(s) + \Phi^-(s), \qquad \Psi(s) = \Psi^+(s) + \Psi^-(s).$$

The pair $\Phi(s)$, $\Psi(s)$ inherits the required properties (i), (ii), and (v). Furthermore,

$$\lambda_1^{-s}\Phi(s) = \lambda_1^{-k/2}(\sqrt{\lambda_1\lambda_2})^{-s}\{\hat{\Phi}(s) + \tilde{\Phi}(s)\}$$

and

$$\lambda_2^{-s}\Psi(s) = \lambda_2^{-k/2}(\sqrt{\lambda_1\lambda_2})^{-s}\{\hat{\Phi}(s) - \tilde{\Phi}(s)\}.$$

These expressions show easily that $\lambda_1^{-s}\Phi(s)$ and $\lambda_2^{-s}\Psi(s)$ are linearly independent, by (7.41) and the nontriviality of $\hat{\Phi}(s)$ and $\tilde{\Phi}(s)$.

The derivation of (iv) from the stricter assumption $k \geq 6q/(q-2)$ is identical with that in (a). This completes the proof of (b).

(c) The proof in this case is the same as in case (b), except that here we apply Theorem 6.3 instead of Theorem 5.6 and Corollary 5.3(a).

(d) Apart from the final statement, the proof is the same as that in case (c), except that Theorem 4.1 replaces Theorem 6.3 in the argument. To prove the assertion made in the last sentence, we invoke Theorem 4.1.

Theorem 4.1 implies the existence of infinitely many linearly independent $\hat{\Phi}(s)$ satisfying the first equality in (7.41) ($\gamma = 1$) and infinitely many linearly independent $\tilde{\Phi}(s)$ satisfying the second equality ($\gamma = -1$). Recall that

$$\Phi(s) = \Phi^+(s) + \Phi^-(s) = \lambda_1^{-k/2}\left(\sqrt{\frac{\lambda_1}{\lambda_2}}\right)^s \{\hat{\Phi}(s) + \tilde{\Phi}(s)\}.$$

(See the proof of part (b) for this.)

Let $t \in \mathbb{Z}^+$ and consider

$$\Phi_j(s) = \lambda_1^{-k/2}\left(\sqrt{\frac{\lambda_1}{\lambda_2}}\right)^s \{\hat{\Phi}_j(s) + \tilde{\Phi}_j(s)\},$$

$1 \leq j \leq t$, with the $\hat{\Phi}_j(s)$ linearly independent, and the $\tilde{\Phi}_j(s)$ as well.

Suppose that

$$\sum_{j=1}^{t} \alpha_j \Phi_j(s) \equiv 0,$$

with complex α_j; that is,

$$\sum_{j=1}^{t} \alpha_j \hat{\Phi}(s) \equiv -\sum_{j=1}^{t} \alpha_j \tilde{\Phi}(s).$$

By (7.41), this implies that

$$\sum_{j=1}^{t} \alpha_j \tilde{\Phi}_j(s) \equiv 0, \quad \text{so} \quad \alpha_1 = \cdots = \alpha_t = 0,$$

and $\Phi_1(s), \ldots, \Phi_t(s)$ are linearly independent over $\mathbb{C}$. Since $t \in \mathbb{Z}^+$ is arbitrary, there are infinitely many linearly independent choices for $\Phi(s)$.

One can apply the same argument to show the existence of infinitely many linearly independent $\Psi(s)$. Alternatively, this follows from the linear independence of the $\Phi(s)$, together with the functional equation (v) relating $\Psi(s)$ with $\Phi(s)$. This concludes the proof. □

Chapter 8

Identities equivalent to the functional equation and to the modular relation

In Chapter 7, we focused our attention on a generalization of Hecke's functional equation due to Bochner. In fact, Bochner [17], [20, pp. 665–696] considered a further more general functional equation than that of Hecke. We begin this chapter by describing the setting for this considerably more general functional equation, although we shall concentrate mainly in the sequel on Hecke's functional equation. Identities arising from other special cases of our general functional equation can be found in [9]. In Chapter 2, we briefly and implicitly alluded to an identity for $\sum_{\lambda_n \leq x} a_n(x-\lambda_n)^\rho$ that is equivalent to Hecke's functional equation. We give not only that identity in this chapter, but offer further identities equivalent to Hecke's functional equation. In Chapters 2 and 7, the Mellin transform of the exponential function and the inverse Mellin transform of the Gamma function play key roles in demonstrating the equivalence of the modular relation and the functional equation. In proving the identities in this chapter, Mellin transforms also play central roles. For example, in establishing an identity for the Riesz sum $\sum_{\lambda_n \leq x} a(n)(x-\lambda_n)^\rho$ in Theorem 8.1, the inverse Mellin transform for the ordinary Bessel function $J_\nu(x)$ of order ν [106, p. 196, Eq. (7.9.1)],

$$J_\nu(x) = \frac{1}{2\pi i}\int_{c-i\infty}^{c+i\infty} \frac{\Gamma(s)}{\Gamma(\nu+1-s)}\left(\frac{x}{2}\right)^{\nu-2s} ds, \qquad 0 < c < \frac{1}{2}\nu + \frac{3}{4},$$

is of prime importance. Versions of Perron's formula are needed to prove Theorem 8.1 as well as the identities for logarithmic sums that are discussed after Riesz sums. Another useful Mellin transform is (8.15) below.

For all of the theorems that we present, specific examples and less general theorems historically preceded the more general theorems. In Definition 5.7 and Example 6.3, we considered, respectively, the Ramanujan tau-function $\tau(n)$ and $r_k(n)$, the number of representations of the positive integer n as a sum of k squares. We focus on these two arithmetical functions to illustrate the theorems that follow. No proofs are given, but sources where proofs may be found are indicated.

In our theorems and examples throughout this section, $s = \sigma + it$, with σ and t both real. Also, C denotes a simple closed curve, or union of simple closed curves, oriented in the positive direction.

Definition 8.1. Let λ_n and μ_n, $1 \leq n < \infty$, be two sequences of positive numbers strictly increasing to ∞, and let $a(n)$ and $b(n)$, $1 \leq n < \infty$, be two sequences of complex numbers not identically zero. Consider the functions $\varphi(s)$ and $\psi(s)$ representable as Dirichlet series

$$\varphi(s) = \sum_{n=1}^{\infty} a(n)\lambda_n^{-s} \qquad \text{and} \qquad \psi(s) = \sum_{n=1}^{\infty} b(n)\mu_n^{-s},$$

with finite abscissas of absolute convergence σ_a and σ_a^*, respectively. If N is a positive integer, let

$$\Delta(s) = \prod_{k=1}^{N} \Gamma(\alpha_k s + \beta_k),$$

where $\alpha_k > 0$ and β_k is complex, $1 \leq k \leq N$. With r real, we say that φ and ψ satisfy the functional equation

$$\Delta(s)\varphi(s) = \Delta(r-s)\psi(r-s) \tag{8.1}$$

if there exists in the s-plane a domain $\mathcal{D}$, which is the exterior of a compact set S, such that in $\mathcal{D}$ a holomorphic function χ exists with the properties:

(i) $\chi(s) = \Delta(s)\varphi(s)$, for $\sigma > \sigma_a$,
$\chi(s) = \Delta(r-s)\psi(r-s)$, for $\sigma < r - \sigma_a*$;

(ii) $\lim_{|t|\to\infty} \chi(\sigma + it) = 0$, uniformly in every interval $-\infty < \sigma_1 \leq \sigma \leq \sigma_2 < \infty$;

(iii) $\chi(s)$ is meromorphic on S.

In particular, if $\Delta(s) = \Gamma(s)$ and if we replace $\varphi(s)$ and $\psi(s)$ by $(2\pi)^{-s}\varphi(s)$ and $(2\pi)^{-s}\psi(s)$, respectively, so that λ_n and μ_n are replaced by $2\pi\lambda_n$ and $2\pi\mu_n$, respectively, then (8.1) reduces to Hecke's functional equation

$$(2\pi)^{-s}\Gamma(s)\varphi(s) = (2\pi)^{-(r-s)}\Gamma(r-s)\psi(r-s). \tag{8.2}$$

We begin with Chandrasekharan and Narasimhan's theorem on the Riesz sums $\sum_{\lambda_n \leq x} a(n)(x-\lambda_n)^\rho$ [23]. A different proof of their theorem was later given by Berndt [14]. We state the theorem in a "weak" form in which the identity holds for $\rho > 2\sigma_a^* - r - \frac{1}{2}$. Under fairly complicated additional hypotheses, the identity is valid for $\rho > 2\sigma_a^* - r - \frac{3}{2}$. For the examples that we give, these additional hypotheses are indeed satisfied, and so we state the examples in a "strong" form. Identities for Riesz sums are important in proving theorems on the average order of arithmetical functions. Even though the identity for $\sum_{\lambda_n \leq x} a(n)(x-\lambda_n)^\rho$ may not be valid for $\rho = 0$, a method using *successive differences* can be employed to obtain results on the average order of $\sum_{\lambda_n \leq x} a(n)$. See, for example, [24], [13], and the several references given in these papers.

Theorem 8.1. *Let $J_\nu(x)$ denote the ordinary Bessel function of order ν, and let $x > 0$ and $\rho > 2\sigma_a^* - r - \frac{1}{2}$. Then the functional equation* (8.2) *implies the identity*

$$\frac{1}{\Gamma(\rho+1)} \sum_{\lambda_n \leq x}{}' a(n)(x-\lambda_n)^\rho$$
$$= \left(\frac{1}{2\pi}\right)^\rho \sum_{n=1}^\infty b(n) \left(\frac{x}{\mu_n}\right)^{(r+\rho)/2} J_{r+\rho}(4\pi\sqrt{\mu_n x}) + Q_\rho(x). \quad (8.3)$$

Here

$$Q_\rho(x) = \frac{1}{2\pi i} \int_C \frac{(2\pi)^s \chi(s) x^{s+\rho}}{\Gamma(\rho+1+s)} ds,$$

where C denotes a closed curve or union of closed curves encircling all of S, and the prime $'$ on the summation sign on the left-hand side of (8.3) *indicates that if $\rho = 0$ and $x = \lambda_n$, for some $n \geq 1$, then only $\frac{1}{2}a(n)$ is counted. Conversely, the identity* (8.3) *implies that* (8.2) *holds.*

As indicated above, in many cases Theorem 8.1 can be strengthened by replacing the condition $\rho > 2\sigma_a^* - r - \frac{1}{2}$ by $\rho > 2\sigma_a^* - r - \frac{3}{2}$; see [23, p. 14, Theorem III] and [14].

A generalization of Theorem 8.1 to the case when $\Delta(s) = \Gamma^m(s)$, where m is a positive integer, has been given by Berndt [7]. To illustrate Theorem 8.1 and the aforementioned extension, we give two examples, both of which are also found in [23].

Example 8.1. Let $r_k(n)$ denote the number of representations of n as a sum of k squares. In the notation of Theorem 8.1 and Definition 8.1,

$a(n) = b(n) = r_k(n)$, $\lambda_n = \mu_n = \frac{1}{2}n$, and $r = \frac{1}{2}k$. Then, if x is replaced by $\frac{1}{2}x$, we find that, for $\rho > \frac{1}{2}(k-3)$,

$$\frac{1}{\Gamma(\rho+1)} \sum_{0 \le n \le x}{}' r_k(n)(x-n)^\rho = \frac{\pi^{k/2} x^{k/2+\rho}}{\Gamma(\rho+1+\frac{1}{2}k)} + \left(\frac{1}{\pi}\right)^\rho \sum_{n=1}^{\infty} r_k(n) \left(\frac{x}{n}\right)^{k/4+\rho/2} J_{k/2+\rho}(2\pi\sqrt{nx}).$$

Example 8.2. Let $\tau(n)$ denote Ramanujan's tau-function. Then $\lambda_n = \mu_n = n$ and $a(n) = b(n) = \tau(n)$, $n \ge 1$. Thus, by Theorem 8.1, for $x > 0$ and $\rho > -\frac{1}{2}$, since $\chi(s)$ is entire in this instance,

$$\frac{1}{\Gamma(\rho+1)} \sum_{n \le x}{}' \tau(n)(x-n)^\rho = \left(\frac{1}{2\pi}\right)^\rho \sum_{n=1}^{\infty} \tau(n) \left(\frac{x}{n}\right)^{6+\rho/2} J_{12+\rho}(4\pi\sqrt{nx}).$$

Certain identities for the *logarithmic sums* $\sum_{\lambda_n \le x} a(n) \log^\rho(x/n)$, where ρ is a nonnegative integer, are also equivalent to the functional equation for Dirichlet series involving $\Delta(s) = \Gamma^m(s)$ [8]. The first general theorems for such sums were proved by Berndt [8], and because these general theorems [8, Theorems 1, 3], even in the case of the simple gamma factor $\Gamma(s)$, are somewhat complicated–in particular, the identities involve integrals of Bessel functions–we content ourselves here with offering three examples for $\rho = 1$. The first examples of this type were derived by A. Oppenheim [89].

Example 8.3. If $d(n)$ denotes the number of positive divisors of the positive integer n and γ denotes Euler's constant, then [8, p. 371], [89]

$$\sum_{n \le x} d(n) \log(x/n) = x(\log x - 2 + 2\gamma) + \frac{1}{4} \log(4\pi^2 x) + \frac{1}{2\pi} \sum_{n=1}^{\infty} \frac{d(n)}{n} \left(Y_0(4\pi\sqrt{nx}) + \frac{2}{\pi} K_0(4\pi\sqrt{nx}) \right),$$

where $Y_\nu(x)$ denotes the second solution of Bessel's differential equation of order ν, and $K_\nu(x)$ denotes the modified Bessel function of order ν, both usually so denoted.

In his paper [89], Oppenheim alludes to an identity involving $r_2(n)$ but does not explicitly state it. C. Müller [83], L. Carlitz [22], Berndt [8], and R. Ayoub and S. Chowla [3], [25, pp. 1189–1191] each proved such an identity, and each used an entirely different method. The identity in question is given by

$$\sum_{n\le x} r_2(n)\log(x/n) = \pi x - \log x + c - \frac{1}{\pi}\sum_{n=1}^{\infty}\frac{r_2(n)}{n}J_0(2\pi\sqrt{nx}). \qquad (8.4)$$

The constant c was given in various forms by the authors cited above and is equal to

$$c = -\log\frac{\Gamma^4(1/4)}{16\pi}, \qquad (8.5)$$

with the best proof probably that of Ayoub and Chowla.

Example 8.4. If $\tau(n)$ denotes Ramanujan's tau-function, then [8, p. 372]

$$\sum_{n\le x}\tau(n)\log(x/n) = 2\left(\frac{1}{4\pi}\right)^{12}\sum_{n=1}^{\infty}\frac{\tau(n)}{n^{12}}\int_0^{4\pi\sqrt{nx}} u^{11}J_{12}(u)du.$$

The next two theorems offer identities for modified zeta functions or Hurwitz-type zeta functions, i.e., identities for Dirichlet series in which the variable n of summation, or more generally λ_n or μ_n, is replaced by $n+a$ (or $\lambda_n + a$ or $\mu_n + a$). The first is due to Chandrasekharan and Narasimhan [23, p. 8, Lemma 6], and the second is due to Berndt [6], [9, p. 342, Theorem 8.1]. In fact, prior to the publication of [23], Bochner and Chandrasekharan [21], [20, 715–739] had established a general identity similar to (8.6) below.

Theorem 8.2. *Let $a(n)$ and $b(n)$, $n\ge 1$, be coefficients of Dirichlet series satisfying the functional equation (8.2). Suppose that ρ is a nonnegative integer. Then if* Re $s>0$, $\sigma_a^*>0$, *and* $\rho > 2\sigma_a^* - r - \frac{1}{2}$,

$$\left(-\frac{1}{s}\frac{d}{ds}\right)^{\rho}\left\{\frac{1}{s}\sum_{n=1}^{\infty}a(n)e^{-s\sqrt{\lambda_n}}\right\}$$
$$= 2^{3r+\rho}\Gamma(r+\rho+\tfrac{1}{2})\pi^{r-1/2}\sum_{n=1}^{\infty}\frac{b(n)}{(s^2+16\pi^2\mu_n)^{r+\rho+1/2}} + R_\rho(s), \qquad (8.6)$$

where

$$R_\rho(s) := \frac{2^{-\rho}}{2\pi i}\int_C \frac{\chi(z)(2\pi)^z\Gamma(2z+2\rho+1)}{\Gamma(z+\rho+1)}s^{-2z-2\rho-1}dz.$$

Here C is a closed curve or union of closed curves containing all poles of the integrand on the interior of C. Conversely, if (8.6) holds for some integral ρ and $\sum_{n=1}^{\infty} |b(n)|\mu_n^{-r-\rho-1/2} < \infty$, then (8.2) is valid.

Example 8.5. If we put $a(n) = b(n) = \tau(n)$ and $\rho = 0$ in Theorem 8.2, then, for Re $s > 0$, we deduce that, since $\chi(s)$ is entire,

$$\sum_{n=1}^{\infty} \tau(n)e^{-s\sqrt{n}} = 2^{36}\pi^{23/2}\Gamma\left(\frac{25}{2}\right)\sum_{n=1}^{\infty} \frac{\tau(n)}{(s^2+16\pi^2 n)^{25/2}}.$$

Theorem 8.3. *Suppose that $\varphi(s)$ satisfies Definition 8.1, with (8.1) taking the simpler form (8.2). Define for $a > 0$ and $\sigma > \sigma_a$,*

$$\varphi(s,a) = \sum_{n=1}^{\infty} a(n)(\lambda_n + a)^{-s}.$$

Let $\mathbb{D}$ be a domain where

$$\sum_{n=1}^{\infty} b(n)K_{s-r}(4\pi\sqrt{\mu_n a})\mu_n^{(s-r)/2}$$

converges uniformly. As above, $K_\nu(z)$ denotes the modified Bessel function of order ν. Let $R(s,a)$ denote the sum of the residues of $\chi(w)\Gamma(s-w)(2\pi a)^{w-s}$ at the poles of $\chi(w)$. Then, if $s \in \mathbb{D}$,

$$(2\pi)^{-s}\Gamma(s)\varphi(s,a) = 2\sum_{n=1}^{\infty} b(n)\left(\frac{\mu_n}{a}\right)^{(s-r)/2} K_{s-r}(4\pi\sqrt{\mu_n a}) + R(s,a). \tag{8.7}$$

Conversely, if $\varphi(s,a)$ satisfies (8.7), then $\varphi(s)$ satisfies (8.2).

It should be emphasized that in the statement of Theorem 8.3, $R(s,a)$ does not include any residues arising from poles of $\Gamma(s-w)$. Keeping this in mind, we can write

$$R(s,a) = \frac{1}{2\pi i}\int_C \chi(w)\Gamma(s-w)(2\pi a)^{w-s}dw, \tag{8.8}$$

where C is a closed curve or union of closed curves enclosing the poles of $\chi(w)$ but not of $\Gamma(w-s)$, unless poles of these two functions coalesce.

If Theorem 8.2 holds for $\rho = 0$, then the theorem coincides with Theorem 8.3 if we set $s = r + \frac{1}{2}$ in Theorem 8.3, set $s = 4\pi^2\sqrt{a}$ in Theorem 8.2, reverse the roles of $\varphi(s)$ and $\psi(s)$, employ the well-known formula [111,

p. 80]

$$K_{1/2}(z) = \sqrt{\frac{\pi}{2z}}e^{-z}, \tag{8.9}$$

and use the representation (8.8). In demonstrating that a common identity is obtained, one must use the functional equation (8.2).

Another proof of Theorem 8.3 was given by S. Kanemitsu, Y. Tanigawa, and M. Yoshimoto [57]. They furthermore show that the identity (8.7) is equivalent to an identity involving incomplete gamma functions due to A. F. Lavrik [71].

Example 8.6. Let $a(n) = \tau(n)$. Then, as in Example 8.2, $b(n) = \tau(n)$, $\mu_n = n$, $r = 12$, and $\chi(w)$ is entire. We find from Theorem 8.3 that, for $a > 0$ and any complex number s,

$$(2\pi)^{-s}\Gamma(s)\varphi(s,a) = 2\sum_{n=1}^{\infty}\tau(n)\left(\frac{n}{a}\right)^{(s-12)/2}K_{s-12}(4\pi\sqrt{na}),$$

where for $\sigma > \frac{13}{2}$,

$$\varphi(s,a) = \sum_{n=1}^{\infty}\frac{\tau(n)}{(n+a)^s}.$$

The following theorem is the special case $\varphi(s) = \zeta(2s)$ of Theorem 8.3 and historically is the first instance of Theorem 8.3 to be established. Here, $r = \frac{1}{2}$ and $s - \frac{1}{2}$ has been replaced by ν, so that we can record the theorem in its original formulation.

Theorem 8.4. *Let $K_\nu(z)$ be as above. If $x > 0$ and $\operatorname{Re}\nu > 0$, then*

$$\begin{aligned}\frac{1}{4}&(\pi x)^{-\nu}\Gamma(\nu) + \sum_{n=1}^{\infty} n^{\nu}K_\nu(2\pi nx)\\ &= \frac{1}{4}\sqrt{\pi}(\pi x)^{-\nu-1}\Gamma(\nu+\tfrac{1}{2}) + \frac{\sqrt{\pi}}{2x}\left(\frac{x}{\pi}\right)^{\nu+1}\Gamma(\nu+\tfrac{1}{2})\sum_{n=1}^{\infty}(n^2+x^2)^{-\nu-1/2}.\end{aligned} \tag{8.10}$$

Theorem 8.4 was first established by G. N. Watson [110], who used the Poisson summation formula. H. Kober [67] generalized Theorem 8.4 in two different directions. Berndt [15] proved a generalization of (8.10) in which either even or odd periodic coefficients appear as coefficients in both infinite series of (8.10). A short, more recent proof of Theorem 8.4 has been given by Kanemitsu, Tanigawa, and Yoshimoto [56].

Recall that the theta function identity

$$\sum_{n=-\infty}^{\infty} e^{-\pi i n^2/\tau} = \sqrt{\tau/i} \sum_{n=-\infty}^{\infty} e^{\pi i n^2 \tau}, \qquad \operatorname{Im} \tau > 0, \tag{8.11}$$

is equivalent to the functional equation of the Riemann zeta function $\zeta(s)$ given by

$$\pi^{-s/2}\Gamma\left(\tfrac{1}{2}s\right)\zeta(s) = \pi^{-(1-s)/2}\Gamma\left(\tfrac{1}{2}(1-s)\right)\zeta(1-s). \tag{8.12}$$

If we consider now the functional equation for $\zeta^2(s)$, then Koshliakov's formula [68] in the theorem below can be considered as an analogue of the transformation formula (8.11) for the classical theta function. To justify this claim, recall first from (2.3) that the exponential function is the inverse Mellin transform of $\Gamma(s)$. Second, recall that the inverse Mellin transform of $2^{s-2}\Gamma^2(\frac{1}{2}s)$ is the modified Bessel function $K_0(x)$ [33, p. 331]. The coefficients of the exponential functions in the classical theta relation (8.11) are 1, while the coefficients of $K_0(2n\alpha)$ in Koshliakov's formula below are $d(n)$, the number of positive divisors of the positive integer n, which are the coefficients in $\zeta^2(s)$. Koshliakov's formula was established in 1929 [68], but, in fact, it was recorded by Ramanujan about ten years earlier and can be found in his lost notebook [91, p. 253].

Theorem 8.5. *Let α and β denote positive numbers such that $\alpha\beta = \pi^2$, let γ denote Euler's constant, and let $d(n)$ denote the number of positive divisors of the positive integer n. Then,*

$$\sqrt{\alpha}\left(\frac{1}{4}\gamma - \frac{1}{4}\log(4\beta) + \sum_{n=1}^{\infty} d(n)K_0(2n\alpha)\right) = \sqrt{\beta}\left(\frac{1}{4}\gamma - \frac{1}{4}\log(4\alpha) + \sum_{n=1}^{\infty} d(n)K_0(2n\beta)\right). \tag{8.13}$$

Koshliakov established Theorem 8.5 by using the Voronoï summation formula [109], which can be considered to be an arithmetic analogue of the classical Poisson summation formula. We do not give a statement of Voronoï's summation formula here, but for several versions of it and conditions under which they are valid, see [11]. A. L. Dixon and W. L. Ferrar [30] also used Voronoï's formula to prove (8.13), while F. Oberhettinger and K. L. Soni [86] established a generalization of (8.13) using Voronoï's formula. Soni [101] derived further identities from Koshliakov's formula. In contrast to the work of these authors, Ramanujan evidently did not appeal to Voronoï's formula. A. P. Guinand [36] showed in 1955 that Koshliakov's formula is a corollary of a more general formula, now called *Guinand's formula*, in Theorem 8.6 below. See also a paper by Berndt, Y. Lee, and J. Sohn [16] for similar, more detailed proofs of both Guinand's and Koshliakov's formulas. We remark that both Guinand [36] and Berndt, Lee, and Sohn [16] employ Theorem 8.4 in their proof of Theorem 8.6. Since Ramanujan recorded Guinand's formula on the same page of his lost notebook [91, p. 253] as he had recorded Koshliakov's formula, he undoubtedly made the same deduction of the latter formula from the former as did Guinand approximately 36 years later. Ramanujan also recorded on page 254 of his lost notebook [91] additional theorems that can be derived from Guinand's formula and that were not later rediscovered by other mathematicians; see [16] for proofs of these results.

Theorem 8.6. *Let $\sigma_k(n) = \sum_{d|n} d^k$. If α and β are positive numbers such that $\alpha\beta = \pi^2$, and if s is any complex number, then*

$$\sqrt{\alpha}\sum_{n=1}^{\infty}\sigma_{-s}(n)n^{s/2}K_{s/2}(2n\alpha)-\sqrt{\beta}\sum_{n=1}^{\infty}\sigma_{-s}(n)n^{s/2}K_{s/2}(2n\beta)$$
$$=\frac{1}{4}\Gamma\left(\frac{s}{2}\right)\zeta(s)\{\beta^{(1-s)/2}-\alpha^{(1-s)/2}\}+\frac{1}{4}\Gamma\left(-\frac{s}{2}\right)\zeta(-s)\{\beta^{(1+s)/2}-\alpha^{(1+s)/2}\}. \tag{8.14}$$

As indicated above, the identity (8.14) was first proved in print by Guinand [36] in 1955. The series in Theorem 8.6 are remindful of the Fourier expansions of nonanalytic Eisenstein series on $SL(2,\mathbb{Z})$, or Maass wave forms [77], [81, pp. 230–232], [69, pp. 15–16], [104, pp. 208–209]. Since certain nonanalytic Eisenstein series, namely, those defined by

$$\sum_{c,d=-\infty}^{\infty}{}' |c\tau+d|^{-s}, \qquad \text{Re } s>2,\ \tau\in\mathcal{H},$$

where the $'$ on the summation sign indicates that the term with $c=d=0$ is omitted, can be recast as Epstein zeta functions, these two objects can be regarded as different sides of the same coin. The Fourier series for nonanalytic Eisenstein series was derived by Maass [77] in the same year, 1949, that A. Selberg and S. Chowla [97], [94, pp. 367–378] published the Fourier series of the Epstein zeta function, but with their proof not published until eighteen years later [98], [94, pp. 521–545]. In the meanwhile, R. A. Rankin [92] and P. T. Bateman and E. Grosswald [5] published proofs. Later proofs were devised by Y. Motohashi [82] and Berndt [12], who generalized the result to more general Epstein zeta functions. In providing another proof of a slightly less general formula, Kanemitsu, Tanigawa, H. Tsukada, and Yoshimoto [54] point out that this less general result was also given by Kober [67, p. 620] in 1934. In fact, Kanemitsu, Tanigawa, Tsukada, and Yoshimoto (or a subset of these authors) have written several papers emphasizing consequences of the modular relation [53], [54], [55], [56], [57], [58]. The nonanalytic Eisenstein series mentioned above were shown by Maass [77] to satisfy modular relations. C. J. Moreno kindly informed the authors that he was easily able to derive Theorem 8.6 from the aforementioned Fourier series expansion and functional equation of the Epstein zeta function. One may then regard (8.14) as an analogue of a modular relation equivalent to the functional equation of these nonholomorphic Eisenstein series or these particular Maass wave forms.

We next state a general theorem in the spirit of Guinand's formula in Theorem 8.6; see [9, p. 343, Theorem 9.1] for a proof. Oberhettinger and Soni [86, p. 24] established a similar theorem a few years later.

Theorem 8.7. *Suppose that $\varphi(s)$ satisfies Definition* 8.1 *in the form* (8.2). *Define, for $x > 0$,*

$$P(x) := \frac{1}{2\pi i}\int_C \chi(s)x^{-s}ds,$$

where C is a closed curve, or union of closed curves, containing the set S of Definition 8.1 *on its interior. If s is any complex number and* $\operatorname{Re} a, \operatorname{Re} b > 0$, *then*

$$\begin{aligned}2\sum_{n=1}^{\infty} a(n)\left(\frac{b}{\lambda_n+a}\right)^{s/2} K_s(4\pi\sqrt{(\lambda_n+a)b}\,)\\ = \int_0^{\infty} x^{s-1}e^{-2\pi ax-2\pi b/x}P(x)dx\\ +2\sum_{n=1}^{\infty} b(n)\left(\frac{a}{\lambda_n+b}\right)^{(r-s)/2} K_{r-s}(4\pi\sqrt{(\mu_n+b)a}\,).\end{aligned}$$

Example 8.7. In Theorem 8.7, set $a(n) = b(n) = r_k(n)$, with $k \geq 2$. In this case, we know that $P(x) = -1 + \pi^{k/2}x^{-k/2}$ [23, p. 19]. We also need the integral evaluation [35, p. 384, formula 3.471, no. 9]

$$\int_0^\infty x^{\nu-1} e^{-\beta/x-\gamma x} dx = 2\left(\frac{\beta}{\gamma}\right)^{\nu/2} K_\nu(2\sqrt{\beta\gamma}), \tag{8.15}$$

where ν is any complex number and $\operatorname{Re}\beta > 0$, $\operatorname{Re}\gamma > 0$. Hence, for any $s \in \mathbb{C}$,

$$\begin{aligned}
&\int_0^\infty x^{s-1} e^{-2\pi a x - 2\pi b/x} P(x) dx \\
&= \int_0^\infty e^{-2\pi a x - 2\pi b/x} \left(-x^{s-1} + \pi^{k/2} x^{s-k/2-1}\right) dx \\
&= -2\left(\frac{b}{a}\right)^{s/2} K_s(4\pi\sqrt{ab}) + 2\left(\frac{b}{a}\right)^{(s-k/2)/2} K_{s-k/2}(4\pi\sqrt{ab}),
\end{aligned} \tag{8.16}$$

where we have made two applications of (8.15). Define $r_k(0) = 1$. In Theorem 8.7, replace a and b by $\frac{1}{2}a$ and $\frac{1}{2}b$, respectively, and use (8.16) to deduce that, for all complex numbers s,

$$\begin{aligned}
&\sum_{n=0}^\infty r_k(n) \left(\frac{b}{n+a}\right)^{s/2} K_s(2\pi\sqrt{(n+a)b}) \\
&\qquad = \sum_{n=0}^\infty r_k(n) \left(\frac{a}{n+b}\right)^{(k/2-s)/2} K_{k/2-s}(2\pi\sqrt{(n+b)a}).
\end{aligned} \tag{8.17}$$

In particular, if $s = \frac{1}{2}$ and $k = 2$, then upon the use of (8.9), we see that (8.17) reduces to the identity

$$\sum_{n=0}^\infty \frac{r_2(n)}{\sqrt{n+a}} e^{-2\pi\sqrt{(n+a)b}} = \sum_{n=0}^\infty \frac{r_2(n)}{\sqrt{n+b}} e^{-2\pi\sqrt{(n+b)a}}. \tag{8.18}$$

This identity was first proved by Ramanujan and communicated to G. H. Hardy, who recorded and sketched a proof of it in his paper [42, p. 283], [44, p. 263] on the famous "circle problem," which we now briefly describe. Let the "error term" $P(x)$ be defined by

$$\sum_{0 \le n \le x}{}' r_2(n) = \pi x + P(x), \tag{8.19}$$

where, as usual, the $\prime$ on the summation sign indicates that if x is a positive integer N, only $\frac{1}{2}r_2(N)$ is counted. In [42], [44, pp. 243–263], Hardy proved that

$$P(x) = \Omega((x\log x)^{1/4}), \tag{8.20}$$

i.e., for *every* positive constant A, there exists a sequence $\{x_n\}$, $n \geq 1$, tending to ∞ such that $|P(x_n)| > A(x_n \log x_n)^{1/4}$, $n \geq 1$. If (8.18) is differentiated with respect to b, and if we then let a tend to 0, we obtain an identity which was crucial for Hardy's proof, and which had earlier been proved by Hardy in a paper written in 1908 [41, p. 373], [45, pp. 434–452, Equation (74), p. 450]. At the end of his paper [42, p. 283], [44, p. 263], Hardy stated the corresponding identity for representations of an integer by an arbitrary positive definite quadratic form in two variables. The identity (8.18) cannot be found in Ramanujan's published papers or notebooks. Another proof of (8.18) was given by Dixon and Ferrar [29]. The symmetry of (8.17) and (8.18) in the parameters a and b is striking.

Example 8.8. Let $a(n) = b(n) = \tau(n)$. We apply Theorem 8.7 and note that $P(x) \equiv 0$ [23, p. 16], [80]. Hence, we find that, for any complex number s,

$$\sum_{n=1}^{\infty} \tau(n)\left(\frac{b}{n+a}\right)^{s/2} K_s(4\pi\sqrt{(n+a)b}) = \sum_{n=1}^{\infty} \tau(n)\left(\frac{a}{n+b}\right)^{(12-s)/2} K_{12-s}(4\pi\sqrt{(n+b)a}). \tag{8.21}$$

To indicate one further consequence of the modular relation, we first need to recall the definition of the Laguerre polynomials $L_n^{(\alpha)}(x)$, namely [35, p. 1061],

$$L_n^{(\alpha)}(x) = \frac{1}{n!}e^x x^{-\alpha}\frac{d^n}{dx^n}(e^{-x}x^{n+\alpha}), \qquad n \geq 0.$$

Theorem 8.8. *Let $L_n^{(\alpha)}(x)$ denote the nth Laguerre polynomial, and let $\varphi(s)$ and $\psi(s)$ satisfy Hecke's functional equation (8.2). Then, if $y > 0$ and m is any nonnegative integer,*

$$\sum_{n=1}^{\infty} a(n)(2\pi\lambda_n)^m e^{-2\pi\lambda_n y} = m!y^{-r-m}\sum_{n=1}^{\infty} b(n)e^{-2\pi\mu_n/y}L_m^{(r-1)}(2\pi\mu_n/y) + \frac{1}{2\pi i}\int_C \frac{\chi(s)\Gamma(s+m)}{\Gamma(s)}y^{-s-m}ds,$$

where C is a closed curve or union of closed curves encircling the poles of $\chi(s)$, and where $-n \notin C$, for $n = 0, 1, \ldots, m$.

Theorem 8.8 is due to Berndt [9, Theorem 10.1] and generalizes a theorem of G. Szegö [103].

In closing this monograph, we indicate one further class of Dirichlet series, axiomatically defined for the first time by A. Selberg [96], [95, pp. 47–63] in 1989. An excellent article describing investigations of this class since 1989 has been written by J. Kaczorowski [51], and for an article addressing the Hecke theory, see [52]. We reproduce the definition of the Selberg class as given in [51].

Definition 8.2. The Selberg class is the set of all Dirichlet series

$$F(s) := \sum_{n=1}^{\infty} \frac{a(n)}{n^s}$$

that satisfy the following axioms.

(1) The Dirichlet series $F(s)$ has abscissa of absolute convergence equal to 1.
(2) *Analytic Continuation.* There exists an integer $m \geq 0$ such that $(s-1)^m F(s)$ is an entire function of finite order.
(3) *Functional Equation.* The function $F(s)$ satisfies a functional equation of the form

$$\Phi(s) = \omega \tilde{\Phi}(1-s),$$

where $\tilde{f}(s) = \overline{f(\overline{s})}$, $|\omega| = 1$, and

$$\Phi(s) = Q^s \prod_{j=1}^{r} \Gamma(\lambda_j s + \mu_j) F(s),$$

where $r \geq 0$, $Q > 0$, $\lambda_j > 0$, and $\operatorname{Re} \mu_j \geq 0$ for $1 \leq j \leq r$, with all parameters depending on $F(s)$.
(4) *Ramanujan Hypothesis.* For every $\epsilon > 0$, $a(n) \ll n^{\epsilon}$.
(5) *Euler Product.* For $\sigma > 1$,

$$\log F(s) = \sum_{n=1}^{\infty} \frac{b(n)}{n^s},$$

where $b(n) = 0$ unless $n = p^m$ with $m \geq 1$, and $b(n) \ll n^{\theta}$ for some $\theta < \frac{1}{2}$.

Observe that, in this definition, Dirichlet series are normalized so that the abscissa of absolute convergence is equal to 1. Dirichlet series in the

Selberg class include the Riemann zeta function, Dirichlet L-functions, and Dedekind zeta functions for algebraic number fields; see [51] for further examples. A major goal in the Selberg class theory is to demonstrate that certain Dirichlet series indeed do belong to the Selberg class; again see [51] for examples.

Bibliography

[1] Ahlfors, L. V. (1966). *Complex analysis*, McGraw-Hill, New York.

[2] Ayoub, R. (1963). *An introduction to the analytic theory of numbers*, American Mathematical Society, Providence, RI.

[3] Ayoub, R. and Chowla, S. (1970). On a theorem of Müller and Carlitz, *J. Number Thy.* **2**, pp. 342–344.

[4] Bateman, P. T. (1951). On the representations of a number as the sum of three squares, *Trans. Amer. Math. Soc.* **71**, pp. 70–101.

[5] Bateman, P. T. and Grosswald, E. (1964). On Epstein's zeta function, *Acta Arith.* **9**, pp. 365–373.

[6] Berndt, B. C. (1967). Generalised Dirichlet series and Hecke's functional equation, *Proc. Edinburgh Math. Soc.* **15**, pp. 309–313.

[7] Berndt, B. C. (1969). Identities involving the coefficients of a class of Dirichlet series. I, *Trans. Amer. Math. Soc.* **137**, pp. 345–359.

[8] Berndt, B. C. (1969). Identities involving the coefficients of a class of Dirichlet series. II, *Trans. Amer. Math. Soc.* **137**, pp. 361–374.

[9] Berndt, B. C. (1969). Identities involving the coefficients of a class of Dirichlet series. III., *Trans. Amer. Math. Soc.* **146**, pp. 323–348.

[10] Berndt, B. C. (1970). Identities involving the coefficients of a class of Dirichlet series. IV., *Trans. Amer. Math. Soc.* **149**, pp. 179–185.

[11] Berndt, B. C. (1971). Identities involving the coefficients of a class of Dirichlet series. V., *Trans. Amer. Math. Soc.* **160**, pp. 139–156.

[12] Berndt, B. C. (1971). Identities involving the coefficients of a class of Dirichlet series. VI., *Trans. Amer. Math. Soc.* **160**, pp. 157–167.

[13] Berndt, B. C. (1971). On the average order of a class of arithmetical functions, I, *J. Number Thy.* **3**, pp. 184–203.

[14] Berndt, B. C. (1975). Identities involving the coefficients of a class of Dirichlet series. VII., *Trans. Amer. Math. Soc.* **201**, pp. 247–261.

[15] Berndt, B. C. (1975). Periodic Bernoulli numbers, summmation formulas and applications, in *Theory and Application of Special Functions* (R. A. Askey, ed.), Academic Press, New York, pp. 143–189.

[16] Berndt, B. C., Lee, Y., and Sohn, J. (2008). Koshliakov's formula and Guinand's formula in Ramanujan's lost notebook, in *Surveys in Number*

Theory, K. Alladi, ed., Developments in Mathematics, Springer, New York, to appear.

[17] Bochner, S. (1951). Some properties of modular relations, *Ann. Math.* **53**, pp. 332–363.

[18] Bochner, S. (1952). Connection between functional equations and modular relations and functions of exponential type, *J. Indian Math. Soc.* **16** pp. 99–102.

[19] Bochner, S. (1958). On Riemann's functional equation with multiple gamma factors, Ann. Math. **67**, pp. 29–41.

[20] Bochner, S. (1991). *Collected papers of Salomon Bochner*, Part 2, American Mathematical Society, Providence, RI.

[21] Bochner, S. and Chandrasekharan, K. (1956). On Riemann's functional equation, Ann. Math. **63**, pp. 336–360.

[22] Carlitz, L. (1957). A formula connected with lattice points in a circle, *Abh. Math. Sem. Univ. Hamburg* **21** (1957), pp. 87–89.

[23] Chandrasekharan, K. and Narasimhan, R. (1961). Hecke's functional equation and arithmetical identities, *Ann. Math.* **74**, pp. 1–23.

[24] Chandrasekharan, K. and Narasimhan, R. (1962). Functional equations with multiple gamma factors and the average order of arithmetical functions, *Ann. Math.* **76**, pp. 93–136.

[25] Chowla, S. (1999). *The collected papers of Sarvadaman Chowla*, Vol. 3, Les Publications Centre de Recherches Mathématiques, Montreal.

[26] Copson, E. T. (1935). *Theory of functions of a complex variable*, Clarendon Press, Oxford.

[27] Davenport, H. (2000). *Multiplicative number theory*, third ed., Springer-Verlag, New York.

[28] Deligne, P. (1974). La conjecture de Weil I, *Publ. I.H.E.S.* **43**, pp. 273–307.

[29] Dixon, A. L. and Ferrar, W. L. (1934). Some summations over the lattice points of a circle (I), *Quart. J. Math.* (Oxford), **5**, pp. 48–63.

[30] Dixon, A. L. and Ferrar, W. L. (1937). On the summation formulae of Voronoï and Poisson, *Quart. J. Math.* (Oxford) **8**, pp. 66–74.

[31] Eichler, M. (1965). Grenzkreisgruppen und kettenbruchartige Algorithmen, *Acta Arith.* **11**, pp. 169–180.

[32] Epstein, P. (1907). Zur Theorie allgemeiner Zetafunktionen. II, *Math. Ann.* **63**, pp. 205–216.

[33] Erdélyi, A., ed. (1954). *Tables of integral tansforms*, Vol. I, McGraw-Hill, New York.

[34] Evans, R. J. (1973). A fundamental region for Hecke's modular groups, *J. Number Thy.* **5**, pp. 108–115.

[35] Gradshteyn, I. S. and Ryzhik, I. M., eds. (1994). *Table of integrals, series, and products*, 5th ed., Academic Press, San Diego.

[36] Guinand, A. P. (1955). Some rapidly convergent series for the Riemann ξ-function, *Quart. J. Math.* (Oxford) **6**, pp. 156–160.

[37] Gunning, R. C. (1962). *Lectures on modular forms*, Princeton University Press, Princeton.

[38] Hamburger, H. (1921). Über die Riemannsche Funktionalgleichung der ζ-Funktion (Erste Mitteilung), *Math. Zeit.* **10**, pp. 240–254.

[39] Hamburger, H. (1922). Über die Riemannsche Funktionalgleichung der ζ-Funktion (Zweite Mitteilung), *Math. Zeit.* **11**, pp. 224–245.

[40] Hamburger, H. (1922). Über die Riemannsche Funktionalgleichung der ζ-Funktion (Dritte Mitteilung), *Math. Zeit.* **13**, pp. 283–311.

[41] Hardy, G. H. (1908), Some multiple integrals, *Quart. J. Math.* (Oxford), **39**, 357–375.

[42] Hardy, G. H. (1915). On the expression of a number as the sum of two squares, *Quart. J. Math.* (Oxford) **46**, pp. 263–283.

[43] Hardy, G. H. (1940). *Ramanujan*, Cambridge University Press, Cambridge; reprinted by Chelsea, New York, 1960; reprinted by the American Mathematical Society, Providence, RI, 1999.

[44] Hardy, G. H. (1967). *Collected papers*, Vol. II, Oxford University Press, Oxford.

[45] Hardy, G. H. (1972). *Collected papers*, Vol. V, Oxford University Press, Oxford.

[46] Hecke, E. (1927). Theorie der Eisensteinschen Reihen höherer Stufe und ihre Anwendung auf Funktionentheorie und Arithmetik, *Hamburger Abh.* **5**, pp. 199–224.

[47] Hecke, E. (1936). Über die Bestimmung Dirichletscher Reihen durch ihre Funktionalgleichung, *Math. Ann.* **112**, pp. 664–669.

[48] Hecke, E. (1938). *Dirichlet series*, Planographed Lecture Notes, Princeton Institute for Advanced Study, Edwards Brothers, Ann Arbor.

[49] Hecke, E. (1944). Herleitung des Euler-Produktes der zetafunktion und einiger L-Reihen aus ihrer Funktionalgleichung, *Math. Ann.* **119**, pp. 266–287.

[50] Hecke, E. (1959). *Mathematische Werke*, Vandenhoeck & Ruprecht, Göttingen.

[51] Kaczorowski, J. (2006). Axiomatic theory of L-functions: the Selberg class, in *Analytic Number Theory* (J. B. Friedlander, D. R. Heath-Brown, H. Iwaniec, and J. Kaczorowski, eds.), Lecture Notes in Math. No. 1891, Springer-Verlag, Berlin-Heidelberg, pp. 133–209.

[52] Kaczorowski, J., Molteni, G., Perelli, A., Steuding, J., and Wolfart, J. (2006). Hecke's theory and the Selberg class, *Func. et Approximatio* **35**, pp. 183–193.

[53] Kanemitsu, S., Tanigawa, Y., and Tsukada, H. (2006), Some number theoretic applications of a general modular relation, *Internat. J. Number Thy.* **2**, pp. 599–615.

[54] Kanemitsu, S., Tanigawa, Y., Tsukada, H., and Yoshimoto, M. (2004). On Bessel series expressions for some lattice sums: II, *J. Phys. A: Math. Gen.* **37**, pp. 719–734.

[55] Kanemitsu, S., Tanigawa, Y., Tsukada, H., and Yoshimoto, M. (2006). Some aspects of the modular relation, in *Proceedings of the Third China–*

Japan Seminar: Tradition and Modernization, Springer, New York, pp. 103–118.

[56] Kanemitsu, S., Tanigawa, Y., and Yoshimoto, M. (2002). On rapidly convergent series for the Riemann zeta-values via the modular relation, *Abh. Math. Sem. Univ. Hamburg* **72**, pp. 187–206.

[57] Kanemitsu, S., Tanigawa, Y., and Yoshimoto, M. (2002). Ramanujan's formula and modular relations, in *Number theoretic methods: future trends* (S. Kanemitsu and C. Jia, eds.), Kluwer, Dordrecht, pp. 159–212.

[58] Kanemitsu, S., Tanigawa, Y., and Yoshimoto, M. (2002). On rapidly convergent series for Dirichlet L-function values via the modular relation, in *Number Theory and Discrete Mathematics*, A. K. Agarwal, B. C. Berndt, C. F. Krattenthaler, G. L. Mullen, K. Ramachandra, and M. Waldschmidt, eds., Hindustan Book Agency, New Delhi, pp. 113–133.

[59] Knopp, M. I. (1966). Polynomial automorphic forms and nondiscontinuous groups, *Trans. Amer. Math. Soc.* **123**, pp. 506–520.

[60] Knopp, M. I. (1993). *Modular functions in analytic number theory*, second ed., Chelsea, New York.

[61] Knopp, M. I. (1994). On Dirichlet series satisfying Riemann's functional equation, *Invent. Math.* **117**, pp. 361–372.

[62] Knopp, M. I. (2000). Hamburger's theorem on $\zeta(s)$ and the abundance principle for Dirichlet series with functional equations, in *Number Theory* (R. P. Bambah, V. C. Dumir, and R. J. Hans-Gill, eds.), Hindustan Book Agency, New Delhi, pp. 201–216.

[63] Knopp, M. I., Lehner, J., and Newman, M. (1965). A bounded automorphic form of dimension zero is constant, *Duke Math. J.* **32**, pp. 457–460.

[64] Knopp, M. I. and Mason, G. (2003). On generalized modular forms, *J. Number Thy.* **99**, pp. 1–28.

[65] Knopp, M. I. and Newman, M. (1993). On groups related to the Hecke groups, *Proc. Amer. Math. Soc.* **119**, pp. 77–80.

[66] Knopp, M. I. and Sheingorn, M. (1996). On Dirichlet series and Hecke triangle groups of infinite volume, *Acta Arith.* **76**, pp. 227–244.

[67] Kober, H. (1934). Transformationsformeln gewisser Besselscher Reihen Beziehungen zu Zeta-functionen, *Math. Z.* **39**, pp. 609–624.

[68] Koshliakov, N. S. (1929). On Voronoï's sum-formula, *Mess. Math.* **58**, pp. 30–32.

[69] Kubota, T. (1973). *Elementary theory of Eisenstein series*, Kodansha, Tokyo.

[70] Landau, E. (1949). *Einführung in die elementare und analytische Theorie der algebraischen Zahlen und der Ideale*, Chelsea, New York.

[71] Lavrik, A. F. (1990). Functional equations with a parameter for zeta-functions (Russian), *Izv. Akad. Nauk SSSR Ser. Mat.* **54**, pp. 501–521; English trans. in *Math. USSR-Izv.* **36** (1991), pp. 519–540.

[72] Lehmer, D. H. (1947). The vanishing of Ramanujan's function $\tau(n)$, *Duke Math. J.* **14**, pp. 429–433.

[73] Lehner, J. (1964). *Discontinuous groups and automorphic functions*, American Mathematical Society, Providence, RI.

[74] Lehner, J. (1966). *A short course in automorphic functions*, Holt, Rinehart and Winston, New York.
[75] Littlewood, J. E. (1944). *Lectures on the theory of functions*, Oxford University Press, New York.
[76] Maass, H. (1949). Über eine neue Art von nichtanalytischen automorphen Funktionen und die Bestimmung Dirichletscher Reihen durch Funktionalgleichungen, *Math. Ann.* **121**, pp. 141–183.
[77] Maass, H. (1964). *Lectures on modular functions of one complex variable*, Tata Institute of Fundamental Research, Bombay.
[78] Milne, S. C. (1996). New infinite families of exact sums of squares formulas, Jacobi elliptic functions, and Ramanujan's tau function, *Proc. Nat. Acad. Sci. USA* **93**, pp. 15004–15008.
[79] Milne, S. C. (2002). Infinite families of exact sums of squares formulas, Jacobi elliptic functions, continued fractions, and Schur functions, *Ramanujan J.* **6**, pp. 7–149; reprinted under the same title as Volume 5 in the Series, *Developments in Mathematics*, Kluwer, Boston, 2002.
[80] Mordell, L. J. (1917). On Mr. Ramanujan's empirical expansions of modular functions, *Proc. Cambridge Philos. Soc.* **19**, pp. 117–124.
[81] Moreno, C. J. (2005). *Advanced analytic number theory: L-Functions*, Math. Surveys and Monographs, Vol. 115, American Mathematical Society, Providence, RI.
[82] Motohashi, Y. (1968). A new proof of the limit formula of Kronecker, *Proc. Japan Acad.* **44**, pp. 614–616.
[83] Müller, C. (1954). Eine Formel der analytischen Zahlentheorie, *Abh. Math. Sem. Univ. Hamburg* **19**, pp. 62–65.
[84] Murty, M. R. (1988). The Ramanujan τ function, in *Ramanujan revisited*, (G. E. Andrews, R. A. Askey, B. C. Berndt, K. G. Ramanathan, and R. A. Rankin, eds.), Academic Press, Boston, pp. 269–288.
[85] Murty, V. K. (1993). Ramanujan and Harish-Chandra, *Math. Intell.* **15**, pp. 33–39.
[86] Oberhettinger, F. and Soni, K. L. (1972). On some relations which are equivalent to functional equations involving the Riemann zeta function, *Math. Z.* **127**, pp. 17–34.
[87] Ogg, A. (1969). *Modular forms and Dirichlet series*, W. A. Benjamin, New York.
[88] Ogg, A. (1969). On modular forms with associated Dirichlet series, *Ann. Math.* **89**, pp. 184–186.
[89] Oppenheim, A. (1927). Some identities in the theory of numbers, *Proc. London Math. Soc.* (2) **26**, 295–350.
[90] Petersson, H. (1949). Über die Berechnung der Skalarprodukte ganzer Modulformen, *Comm. Math. Helv.* **22**, pp. 168–199.
[91] Ramanujan, S. (1988). *The lost notebook and other unpublished papers*, Narosa, New Delhi.
[92] Rankin, R. A. (1953). A minimum problem for the Epstein zeta function, *Proc. Glasgow Math. Assoc.* **1**, pp. 149–158.
[93] Rankin, R. A. (1988). The Ramanujan τ function, in *Ramanujan revisited*,

(G. E. Andrews, R. A. Askey, B. C. Berndt, K. G. Ramanathan, and R. A. Rankin, eds.), Academic Press, Boston, pp. 245–268.

[94] Selberg, A. (1980). *Collected papers*, Vol. I, Springer-Verlag, Berlin.

[95] Selberg, A. (1991). *Collected papers*, Vol. II, Springer-Verlag, Berlin.

[96] Selberg, A. (1992). Old and new conjectures and results about a class of Dirichlet series, *Proc. Amalfi conference on analytic number theory* (*Maiori,* 1989), (E. Bombieri, et al., eds.), Università di Salerno, pp. 367–385.

[97] Selberg, A. and Chowla, S. (1949). On Epstein's zeta-function (I), *Proc. Nat. Acad. Sci.* (USA) **35**, pp. 371–374.

[98] Selberg, A. and Chowla, S. (1967). On Epstein's zeta-function, *J. Reine Angew. Math.* **227**, pp. 86–110.

[99] Siegel, C. L. (1922). Bemerkungen zu einem Satz von Hamburger über die Funktionalgleichung der Riemannschen Zetafunktion, *Math. Ann.* **86**, 276–279.

[100] Siegel, C. L. (1966). *Gesammelte Abhandlungen*, Band I, Springer-Verlag, Berlin.

[101] Soni, K. (1966). Some relations associated with an extension of Koshliakov's formula, *Proc. Amer. Math. Soc.* **17**, pp. 543–551.

[102] Swinnerton-Dyer, H. P. F. (1988). Congruence properties of $\tau(n)$, in *Ramanujan Revisited*, (G. E. Andrews, R. A. Askey, B. C. Berndt, K. G. Ramanathan, and R. A. Rankin, eds.), Academic Press, Boston, pp. 289–311.

[103] Szegö, G. (1926). Beiträge zur Theorie der Laguerreschen Polynome. II. Zahlentheoretische Anwendungen, *Math. Z.* **25**, pp. 388–404.

[104] Terras, A. (1985). *Harmonic analysis on symmetric spaces and applications* I, Springer-Verlag, New York.

[105] Titchmarsh, E. C. (1939). *The theory of functions*, 2nd. ed., Oxford University Press, Oxford.

[106] Titchmarsh, E. C. (1948). *Theory of Fourier integrals*, 2nd ed., Clarendon Press, Oxford.

[107] Titchmarsh, E. C. (1951). *The theory of the Riemann zeta-function*, Clarendon Press, Oxford.

[108] van der Bliji, F. (1950). The function $\tau(n)$ of S. Ramanujan, *Math. Student* **18**, pp. 83–99.

[109] Voronoï, M. G. (1904). Sur une fonction transcendante et ses applications à la sommation de quelques séries, *Ann. École Norm. Sup.* (3) **21**, pp. 207–267, 459–533.

[110] Watson, G. N. (1931). Some self-reciprocal functions, *Quart. J. Math.* (Oxford) **2**, pp. 298–309.

[111] Watson, G. N. (1966). *Theory of Bessel functions*, 2nd ed., University Press, Cambridge.

[112] Weil, A. (1967). Über die Bestimmung Dirichletscher Reihen durch Funktionalgleichungen, *Math. Ann.* **168**, pp. 149–156.

[113] Weil, A. (1977). Remarks on Hecke's lemma and its use, in *Algebraic Number Theory* (S. Iyanaga, ed.), Japan Society for the Promotion of Science, pp. 267–274.

[114] Weil, A. (1979). *Collected papers*, Vol. III, Springer-Verlag, New York.

Index